The Aquarium
A Complete Guide

The Aquarium
A Complete Guide

Freshwater and Marine Fish
Aquaria Care
Equipment and Maintenance

Angelo Mojetta

BLANDFORD

NOTE

Where practical alternatives can be offered in imperial equivalents, dual measurements (metric/imperial) have been given. Liters have been converted to the British gallon (1 gal = 4.55 l) and kilograms to the avoirdupois pound (1 lb = 0.453 kg).

A BLANDFORD BOOK

First published in Great Britain in 1993 by Blandford
An imprint of Cassell
Cassell plc
Wellington House
125 Strand
London WC2R 0BB
Reprinted 1995

Copyright © 1993 Arnoldo Mondadori Editore S.p.A., Milan
English translation copyright © Arnoldo Mondadori Editore S.p.A., Milan

Fish entries and Families section by Donald Wilkie
Symbols by Giovanna Carturan
Colour drawings of fish and invertebrates by Amedeo Gigli and Egidio Imperi
Black and white drawings and drawings of plants by Paolo Rondini

English translation by John Gilbert

All rights reserved. No part of this book may be reproduced or transmitted in any form or by any means electronic or mechanical, including photocopying, recording or any information storage and retrieval system, without prior permission in writing from the copyright holder and Publisher.

ISBN 0–7137–2398–X

A CIP catalogue record for this book is available from the British Library

Printed and bound in Spain by Artes Graficas Toledo, S.A.
D.L : TO-688-1995

CONTENTS

PREFACE	7
INTRODUCTION	8
The aquarium: origins and functions	10
Tanks: materials, dimensions, structures	12
Water: the raw material	15
Heating and lighting	18
Aeration and filtration	21
Equipment and accessories	25
Preparing the aquarium	28
Aquarium maintenance	33
TROPICAL AND TEMPERATE FRESH WATERS	35
TROPICAL SEA WATERS	65
TEMPERATE SEA WATERS	111
PLANTS AND INVERTEBRATES	123
APPENDIX	133
Families	135
Disease guide	152
Glossary	158
Index	160
Bibliography	164

LEVEL OF DIFFICULTY

 easy

 moderate

 difficult

BEHAVIOUR

 aggressive

 social

POSITION IN AQUARIUM

 top

 middle

 bottom

AQUARIUM TYPE

 single species

 community

PREFACE

This book is designed to provide aquarists, both newcomers and experts, with a guide to the range of fish it is possible to keep in an aquarium. All the essential information is given on the keeping of both marine and freshwater species. For this reason, the book is structured according to the environments in which the fish live: temperate and tropical fresh waters; tropical sea waters, temperate sea waters.

A glance at the gamut of information on the ecology of fish will show how vast the subject is. It can be very hard to establish categories and divisions that are clear-cut and not at all blurred. In our method of categorization we are aware that the labelling of fish as "tropical" and "temperate" is neither an ideal one or an easy one. Compromises have been made, which at the same time have taken for granted the existence of tropical, subtropical, warm temperate and cool temperate waters, without being able to define the actual confines of the environments. Despite difficulties in fitting all entries into acceptable slots, we believe that once the reader has established from the Introduction *the principles of an aquarium, they will be able to make an intelligent selection from this representation of aquarium fish.*

Species are listed alphabetically, using the Latin names. This avoids any confusion between species, but also allows a grouping of those belonging to the same genera. It is often the case, furthermore, that the Latin names are used as a convention among aquarists. A systematic listing, while scientifically accurate, would have made it too complicated for an aquarist unfamiliar with notions of taxonomy to find a particular species with any speed.

Each fish is described according to its dimensions, feeding habits and requirements, aquarium needs and any information particular to the fish regarding its biology and behaviour. The appendix also outlines general characteristics of the single families, listing the species included in the volume underneath.

As the book is mainly dedicated to aquarium fish, the part dealing with plants and invertebrates, while meriting a chapter of its own, has been somewhat limited. However, throughout the text there is mention of the importance of these aspects in the tank. We hope that this information will lead the reader to satisfy any further curiosity in additional literature specifically on the subject, thereby completing their aquarium reading.

INTRODUCTION

The aquarium: origins and functions

The tank: materials, dimensions, construction

Water: the raw material

Heating and lighting

Aeration and filtration

Equipment and accessories

Preparing the aquarium

Aquarium maintenance

The aquarium: origins and functions

Approximately 70 per cent of our world is covered by water. Our own bodies contain roughly 70 per cent of the same substance. Considering these simple facts, it is hardly surprising that people have always been fascinated by the watery element, and, furthermore, that this interest should have reached new heights in the twentieth century, in which mankind has expressed an unprecedented concern for the world and its natural environments. However, we do not know exactly when and where the interest in keeping an aquarium originated. Fishing, of course, was one of mankind's earliest activities, and while the main purpose of this occupation was to procure food, there is no reason not to assume that our distant ancestors also found it enjoyable and relaxing, as well as useful, to observe fish swimming around in clear, unpolluted pools, streams and bays.

In historic times, it would appear that as long as 4,500 years or so ago, the ancient Sumerians bred fish in enclosed pools. The Egyptians, in turn, honoured and embalmed certain fish of the Nile, and the Romans attributed divine qualities to moray eels, some wealthy citizens exhibiting them in their villas, in tanks connected to the sea. In China, around 1135 B.C., there already existed precise rules relating to the construction of pools for the rearing of carp. These regulations were collated, in about 460 B.C., for publication in a genuine treatise on aquaculture by Fan Lee, a celebrated politician and administrator of the time.

The history of fish-keeping, not always easy to distinguish from aquaculture, or fish-farming, therefore had its origins in China, where it was associated principally with the goldfish. Towards the end of the twelfth century, this species was kept widely as a pet, so much so that emperors employed specialized craftsmen to build tanks of the most translucent porcelain in which to display the fish in all their variegated colours. In 1596 a certain Chan Ch'ien-te wrote *The Book of the Goldfish (Chu sha yü p'u)*, the first real handbook of fish-keeping, in which he gave instructions on how to feed goldfish, how to breed them and how to look after them, entailing changing the water in which they lived and using a siphon to cleanse the bottom of the tank of detritus.

In Europe, it was towards the end of the seventeenth century that the fashion of keeping fish in the home was introduced, possibly pioneered by the famous Madame de Pompadour. According to consular sources, the two favourite ornamental fish were the goldfish (*Carassius auratus*) and the paradise fish (*Macropodus opercularis*). In London, as Samuel Pepys reported in his *Diary*, many nobles kept coloured fish in glass tanks.

The word "aquarium," in the sense of a water container for keeping and displaying animals and plants, seems to have been used for the first time in a book entitled *A Naturalist's Rambles on the Devonshire Coast* by Philip Gosse, published in 1853. Gosse was responsible, though not solely, for the opening of the first public aquarium at the London Zoo in that same year. Actually, it was in 1819 that W.T. Brande first devised the principles of the balanced aquarium, in which plants and fish could live together in equilibrium. This idea soon became the object of detailed study, and in 1850 the chemist Robert Warrington read a paper to the Chemical Society of London in which he described his own experiments in raising fish in glass tanks "without the need to change the water even for the most demanding plants." It was along these lines that aquaria were to be built for more than half a century to come.

Since then, the popularity of fish-keeping has spread far and wide. Apart from the multitude of visitors to large public aquaria, there are millions of people all over the world who have found it a rewarding and fascinating hobby to be pursued at home. Further to simple domestic fish-keeping, science has developed its use of aquaria, adopting them as virtually indispensable instruments of study in investigating the biology, physiology and behaviour of a multitude of aquatic organisms, as well as in the fields of pharmacological and medical research. From the first simple aquaria containing a few sturdy, resistant plants and fish, we have progressed to large structures that are sometimes incredibly sophisticated, where a range of natural watery habits can be precisely reproduced so that delicate fish and invertebrates can not only be kept for years but also induced to breed.

The aquarium, therefore, has an important educational function to play, particularly in its public form. Apart from the entertainment value, it furnishes information and helps to spread awareness and alter attitudes, primarily in making every one of us aware of our responsibilities towards the natural world and of our own place in it.

Right: a shoal of discus fish.

Below: goldfish on a Japanese engraving from Tansyu.

Left: the Naples Aquarium, founded in 1872.

The tank: materials, dimensions, construction

For the aquarist, the first indispensable step is to select a tank. The correct choice of a tank will guarantee the success of all future operations and will provide a happy introduction to the world of fish-keeping. At one time the only tanks available were made of glass, with a metal frame. They were strong but they required very careful maintenance, especially if they were to be filled with sea water, which would soon begin to corrode the metal. This was a time when silicone adhesives were not widely available, so that the retaining capacity of the glass depended upon putty and mastic, often prepared by amateurs. Consequently, in time they lost their original elasticity and waterproofing qualities, particularly if the aquarium, for any reason, had to be emptied or cleaned.

The introduction and diffusion of silicone adhesives proved very beneficial to aquarists, for tanks could now be made in a variety of forms and sizes simply by cutting and shaping panes of glass suited to the purpose. This gradually gave way to tanks made from a single piece of acrylic material. At first these had certain disadvantages, for they became progressively opaque, scratched easily and were available in only a few sizes. Modern materials, however, are highly reliable, as is evidenced by their use in a number of recently constructed public aquaria (e.g. Baltimore, Osaka, La Rochelle and Genoa), where the tanks, for additional reasons of weight and security, are made of acrylic resin.

There are many other options as well. In cases where it is unnecessary to see into the tank from every side, it is possible to use plainer materials such as cement, or even unusual ones like wood, provided this is plywood stuck together with waterproof adhesives, subsequently reinforced with epoxy resin or nontoxic polyester. One could even make limited use of large plastic food containers. Available in various sizes, they are easily transformed into aquaria by replacing one of the sides with a glued pane of glass. In theory an aquarium can also be made out of a large, stout polythene bag of the type used for breeding microorganisms in aquaculture.

Having examined these alternatives, you must decide upon the one that is most practical and most suited to your needs. Before buying or building a tank, you have to reflect on two things. Firstly, the aquarium, particularly if it is large, will become an integral part of the furniture and, once in place, will be virtually unmovable. Secondly, the aquarium is intended mainly to accommodate living creatures and therefore any personal considerations must be subordinated to the needs of those occupants. The shape of the tank should be determined not only by aesthetic choice, but also by certain other considerations. For example, no two aquaria, even of identical volume, will contain the same number of fish and a narrow, tall tank will accommodate fewer fish than a broad tank as deep as it is high. The former shape, however, not only guarantees the fish better living conditions, but also provides a better overall view because, as a result of refraction, the breadth and depth of the water appear to be reduced.

SITING THE TANK

Interior designers and psychologists agree that it is a good idea to install an aquarium in a dentist's waiting room or even in the surgery itself as a means of relaxing patients. The same principle seems to apply to less stressful circumstances such as the living room of the home, but it is not easy to lay down hard and fast rules as to where a fish tank should be located, there being so many variables involved, not least individual taste.

It is far simpler, therefore, to advise where not to situate the aquarium. Fish and aquatic organisms live in a sensory world in which sounds, however slight, are extremely important. For this reason tanks should be placed as far away as possible from vibrations, such as those caused by machines and, if possible (although harmful effects have not yet been proven), well away from stereo equipment.

Because water is the best of all solvents, and an aquarium is not isolated from its surroundings but directly linked to them through aerators, it is best not to keep tanks in the kitchen or in smoky places, nor to use spray products excessively. Avoid, as far as possible, a location too close to a window. Although the influence of direct sunlight may benefit the coloration and breeding habits of certain fish, and the growth rate of plants, it is likely to cause overheating in summer as well as a proliferation of ugly algae. It is better to situate aquaria where they can receive indirect lighting, enabling the fish to adapt gradually to the light after the darkness of night before being exposed once more to the interior illumination of the aquarium.

The ideal place will, therefore, be one where light and temperature factors are as uniform as possible. In addition, it should be near an electricity point so that all necessary wiring associated with aquarium maintenance can conveniently be connected without being too long and unsightly. Sufficient room should be allowed for such maintenance. Often, for aesthetic reasons, tanks are fitted so tightly into a recess, flanked by shelves or cupboards, that there is no way of getting at them without dismantling them. Nor should aquaria be placed too close to expensive carpets or furniture; there is always a chance that sooner or later, in the course of the everyday maintenance, water will splash out, perhaps with distressing consequences.

Always make sure that the point in the room where you decide to position the aquarium, as well as the structure on which you place it, is sufficiently strong to bear the weight. Ordinary floors, especially away from the walls towards the center of the

False plastic backing gives additional depth to the aquarium.

All sorts of items can be used to create shelters and hiding places.

Wood and cork are particularly suitable for interior furnishing.

Plants are also available in small pots for transplanting.

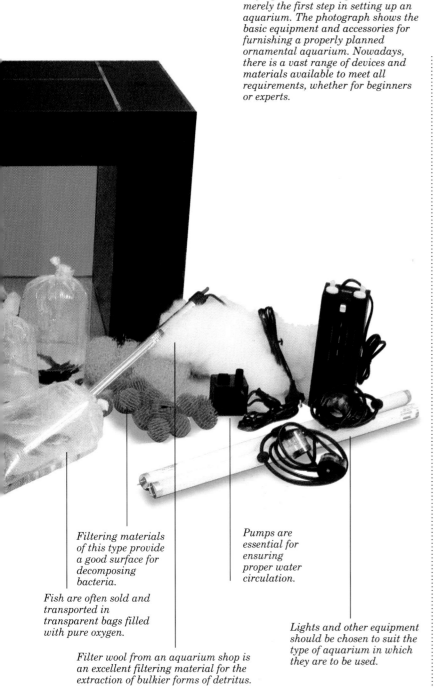

Buying or constructing a tank is merely the first step in setting up an aquarium. The photograph shows the basic equipment and accessories for furnishing a properly planned ornamental aquarium. Nowadays, there is a vast range of devices and materials available to meet all requirements, whether for beginners or experts.

Filtering materials of this type provide a good surface for decomposing bacteria.

Fish are often sold and transported in transparent bags filled with pure oxygen.

Filter wool from an aquarium shop is an excellent filtering material for the extraction of bulkier forms of detritus.

Pumps are essential for ensuring proper water circulation.

Lights and other equipment should be chosen to suit the type of aquarium in which they are to be used.

AQUARIUM CONSTRUCTION

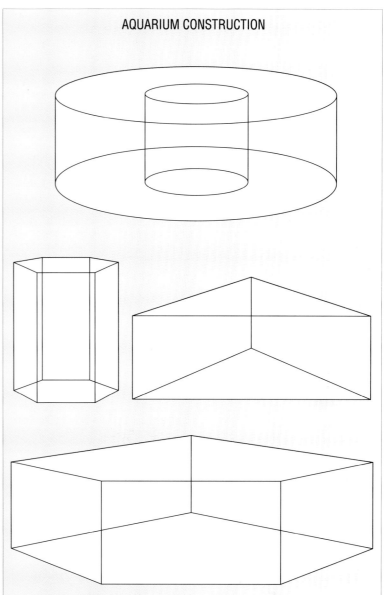

room, are not designed to support really heavy loads. Particularly with large aquaria, therefore, it is worth consulting an expert beforehand. Do not forget that every liter (1¾ pints) of water weighs a kilogram (2¼ lb), and that to the weight of water you have to add approximately 20 per cent for the weight of the tank and the articles inside. So even a small 50-liter (11-gallon) tank, when fitted out, may weigh more than 60 kg (132 lb); and since this is proportional to the volume, loads of 200 – 300 kg (440 – 660 lb) and more may have to be accommodated, so that it is vitally important to make certain that all supports are adequate.

It should be obvious, then, why it is almost obligatory to set up the tank in its permanent position. Any attempt to move it will certainly risk breakage. It is also essential to cushion the tank on a layer of extruded polystyrene at least 1 cm (½ in) thick. This will not only provide a thermal seal to the bottom of the tank and deaden slight vibrations, but will also ensure that any unevenness of the surface, which can be checked with a spirit level, will not exert pressure at key points and possibly crack the base of the tank.

In the past, construction and design of an aquarium was limited by the strength and durability of the materials available. Nowadays, however, with the introduction of good, clear glass and strong, flexible silicone adhesives, all-glass tanks can be constructed in the shape desired. Round, triangular or even hexagonal tanks can be designed to fit into the interior of a room, perhaps to fill an angle otherwise not utilized. As well as glass, there is now also acrylic resin materials, such as Plexiglas, that are visually perfect and weigh a good deal less than glass counterparts. This makes them practically indispensable for very large tanks, for example with sheets over $10m^2$, Nevertheless, glass remains the most popular choice for ornamental aquaria, not only because it is relatively cheap and readily available, but also because of its sturdiness and resistance to handling and scratching. Unfortunately, the same solidness also means that eventual scratches cannot be removed, while marks left on acrylic materials can be eliminated, leaving the quality unchanged over time. Another significant difference, which is likely to bring more application of these plastic materials in future aquarium construction, is the superior quality of thicker sheets. Thick pieces of glass have an intrinsic greenness to them that alters vision of colours, while acrylic sheets are completely clear and colourless. Furthermore, the latter are proving to be more water-tight.

The construction of the tank begins with the positioning of the base sheet (1) and the gluing of the first side sheet (2). The next side sheets should be positioned and glued one at a time (3,4). Below: an aquarium designed to fit into the surroundings.

CHOICE OF TANK

First of all, it is worth ignoring the still widespread belief that a small tank is more suitable than a large one for its occupants. The chemical and physical properties of the water in a large tank are more stable and easily controlled than those in a small tank. The fish will have more room, and there are better opportunities for plant arrangements. Small tanks are therefore best left to expert aquarists, some of whom manage to get excellent results despite the restrictions in volume.

As a general rule, it is advisable to buy a tank measuring about 60 cm (24 in) long x 30 cm (12 in) wide x 38 cm (15 in) high. These dimensions, applicable to freshwater aquaria, should be increased somewhat for saltwater aquaria, to 90 cm (36 in) x 35 cm (14 in) x 45 cm (18 in). Nowadays, there are ready-made tanks on the market in a wide range of sizes up to 500 liters (110 gal), so there should be no problem finding a suitable one. There is also the possibility of having a tank made to measure or in a particular shape according to your specification. Alternatively, you can try constructing it yourself, provided some simple rules are observed, the most important of which is not to be in too much of a hurry. The principal advantages of a home-made aquarium is that you can choose the most suitable size and shape, and also have the satisfaction of saving money.

To build an aquarium, start with a pane of half crystal or of plate glass. In any case, the glass must be perfect from the optical viewpoint. There is nothing worse than looking into an aquarium through distorted glass. As a rule, the necessary panes should be of a minimum thickness of 6 mm (¼ in) for tanks 60 cm (24 in) long and 30 – 40 cm (12 – 16 in) high. The thickness will be increased by 2 mm for every 30 cm (12 in) of additional length. In tanks measuring over 100 cm (39 in), it is advisable to insert cross supports near the top so as to avoid the typical outward curvature of glass because of the pressure exerted by the water on the walls. Particular attention must be paid to the dimensions of the side panes, bearing in mind the thickness of the panes themselves when calculating measurements.

Gluing is done by spreading a uniform layer of silicone, usually pale, translucent in colour. Be careful to use only specific adhesives for aquaria, inasmuch as some silicone adhesives contain additives that may prove toxic for aquatic organisms. When putting everything together, use a table or other work surface that you can walk around at each successive stage to get a good look at the aquarium from every angle. Clean all the panes thoroughly and then start with the base, attaching one front and one side pane so as to form a trihedron. Before the silicone dries completely, check that the panes are perfectly aligned and right-angled, so as to avoid possible trouble later. Before gluing together the remaining parts, make certain the silicone has dried. To prevent the glue smearing all over the glass, line the edges with adhesive tape. When stripped off later, it will take any glue that may have oozed out; any remaining can be removed with a penknife.

When you have finished and you are sure that the silicone has hardened, test the strength of your tank, first supporting it on a layer of polystyrene. Fill it with water gradually, leaving time to check on eventual leaks and bearing in mind that these can still appear hours afterwards.

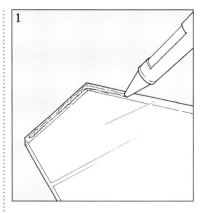
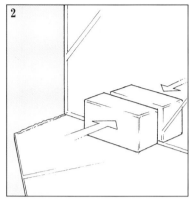
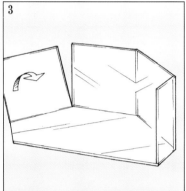
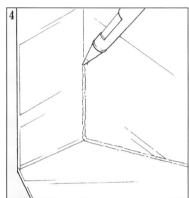

Water: the raw material

The clear waters of a temperate river or stream offer plenty of inspiration for planning an aquarium on a particular theme.

Water, of course, is the vital element. Biologists agree that life began in water, and fish, along with thousands of other creatures, are inextricably linked with it, as shown by their form, their structure and their biology. For chemists, on the other hand, water is a compound derived from the combination of oxygen and hydrogen. This basic liquid, however, is also responsible for the considerable variations that exist among aquatic environments. If we follow the different stages in the water cycle and examine more closely what it contains (water is considered the best possible solvent), we find that every aquatic habitat has its own chemical as well as physical characteristics. Thus a mountain stream will have a lesser concentration of salts than the lower courses of the river formed from it; and in a single region, such as the Amazon, we may find, only a few miles away from each other, waters that are poor in salts and others that are much richer, and waters that are variously acid or alkaline or even brackish, as in certain African lakes. The differences are no less apparent in sea water. From the mouths of rivers to the waters circulating around coral reefs, there is a progressive increase in salt concentration, on the basis of which it is possible to distinguish and classify them accordingly, from the estuarine water, low in salt, through to sea water proper, with a salt content of 30 – 40 per cent and more (*see* table of water classification). In this context it is worth pointing out how marine organisms are better able to withstand internal variations, caused by phenomena of dilution or concentration, rather than external variations in the chemical composition of the water. This is why we have to add, at regular intervals, trace elements, which in an enclosed and restricted environment such as an aquarium, tend inevitably to be broken down and depleted.

PRINCIPAL MINERAL SUBSTANCES DISSOLVED IN FRESH WATER. VALUES ARE AVERAGE AND EXPRESSED IN PERCENTAGES

	Ca^{2+}	Mg^{2+}	Na^+	K^+	CO_3^{2-}	SO_4^{2-}	Cl^-	SiO_4^{2-}
Fresh waters (world average)	63.5	17.4	15.7	3.4	73.9	16.0	10.1	
Rivers (world average)	52.6	24.0	19.2	4.1	67.1	16.3	15.4	
Central Europe	68.2	25.4	4.5	1.9	85.2	10.8	3.9	
Sweden	67.3	16.9	13.6	2.2	74.3	16.2	9.5	
Lake Baikal	59.0	27.0	9.4	4.7	84.0	8.0	1.6	(6.0)
Wisconsin	46.9	37.7	10.9	4.8	69.6	20.5	9.9	
Sierra Nevada	52.0	14.0	26.0	8.0	89.0	5.0	3.0	(4.1)
Germany	36.0	14.3	43.0	6.7	42.4	14.1	43.5	
Spain	38.2	27.0	31.4	3.6	59.8	22.2	14.3	(3.8)
Canada	39.0	37.0	19.0	5.0	61.3	24.5	14.2	
River Congo	43.6	31.6	20.5	4.3	85.0	5.6	9.3	
River Amazon	57.0	17.0	22.0	4.0				
Lake Tanganyika	8.1	44.3	38.2	0.1	83.5	1.2	9.3	

CHEMICAL AND PHYSICAL PROPERTIES

As already mentioned, water is the best of all known solvents. This compound can dissolve a large number of substances, both solid and gaseous. It is this characteristic that is largely of concern to the aquarist and will determine what we need to examine.

pH

The term pH, followed by a number, defines the acidity or alkalinity of water or liquid. To give an example, common vinegar or lemon juice are acids whereas a solution of bicarbonate of soda is alkaline. By convention, the pH is measured by means of a scale of numbers from 1 to 14. The intermediate value 7 corresponds to neutral water. Lower values indicate an ever-increasing acidity. Water of pH 3 will be ten times more acidic than water of pH 4. Values above 7 are typical of alkaline waters, up to a maximum of 14. In nature acid and alkaline waters are likely to be encountered within a more restricted range than this scale indicates.

Water that is too acid or too alkaline is in fact incompatible with the life of fish and other aquatic organisms, whether animal or vegetable. The greatest variation in this parameter, very easy to measure, occurs in fresh water. Here we may come across typically acidic water (pH 5 – 5.6), like that of some rivers in Thailand or certain rivers of the Amazon basin, such as Rio Negro or Rio Tapajoz, or markedly alkaline water as a result of high evaporation and large amounts of dissolved salts, as in some African lakes. The waters of Lake Tanganyika, for example, reach pH 8.5 and those of Lake Magadi, to which the cichlid *Tilapia grahami* has adapted, an exceptional pH 10.5. The pH of sea water is less variable, due to the characteristics of its many

dissolved salts; on average it fluctuates between 8.1 and 8.5. It is vital to consider the pH content of the water if you wish to keep fish and, more especially, if you are trying to breed them.

Hardness

Thanks to its solvent capacity, water that flows over ground or is in contact with it picks up many substances that are subsequently dissolved. The greater or lesser amounts of some of these substances determines whether the water is hard or soft. In fact, the term "hardness" indicates the quantity of salts of calcium and magnesium, these being the more important but not the only ones. Moreover, there are two types of hardness: permanent and temporary. The former (GH or general hardness) depends on the quantity of combined calcium and magnesium in the form of chlorides, sulphates, phosphates, etc. The latter (KH or carbonate hardness) depends on the bicarbonate of calcium present, which can be eliminated by boiling the water. Permanent hardness may be eliminated, too, by more complex chemical methods such as distillation or by chemical processes involving resins and ion exchange. As a general rule, acidic water is soft. This parameter can be checked by simple tests that will be described in due course.

Hardness is usually measured in German degrees identified by the initials DH. Each degree corresponds to a quantity of carbonate of calcium ($CaCO_3$), equivalent to 17.9 ppm, i.e. 17.9 mg per liter. This is a highly important parameter for on it depends the life of fish, which are particularly sensitive to this environmental features. Many fish that lay eggs prefer soft water because an excessive amount of salt would interfere with egg development by hindering indispensable exchanges of gases and salts. The same is true of plants, whose apparent incompatibility is simply a different response to an environment which is favourable to some and unfavourable to others. In addition to the aforementioned salts, a detailed analysis of both fresh water and sea water reveals other substances that are often present in infinitesimal amounts, measurable only by high-precision instruments. For example, in sea water it is possible to detect all the known natural elements as well as those produced by man and responsible for the most serious forms of pollution. Almost the same is true of fresh water, although it shows greater variability, depending fundamentally on the characteristics

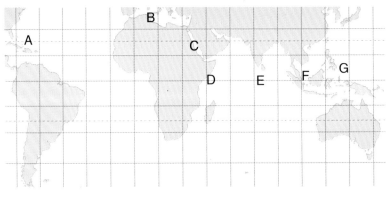

AVERAGE SALT CONCENTRATIONS OF SEAS FROM WHICH COME THE MOST IMPORTANT AQUARIUM SPECIES IN GRAMMES PER LITER

A = Caribbean Sea	35 g/l
B = Mediterranean Sea	37/38 g/l
C = Red Sea	40 g/l
D = Seychelles	32/35 g/l
E = Sri Lanka – Maldive Islands	30/34 g/l
F = Singapore Coastline	30 – 32 g/l
G = Philipinnes	30 – 34 g/l

WATER HARDNESS

mg/l $CaCO_3$	DH	Water type
0 – 50	0 – 3	very soft
50 – 100	3 – 6	soft
100 – 200	6 – 12	moderately hard
200 – 300	12 – 18	hard
300 – 450	18 – 25	very hard
>450	>25	very hard

1 DH = 17.9 mg/l $CaCO_3$

Opposite: control of the chemicals present in water is essential for the purpose of stocking especially demanding fish and plants.

Below: the unusual prism shape of this aquarium is an elegant way of filling the corner of a room.

of the rocks and soils with which it comes into contact. One can recognize waters of different origins by tasting them; water that flows over soil containing iron will tend to taste "ferrous" or "rusty," whereas other waters have a sulphurous taste. To understand how the chemical composition of water can vary, it is enough to read the chemical analysis on the labels of mineral waters.

It is clear that the concentrations of some dissolved substances alter enormously from one place to another. Therefore, when filling the aquarium, given the variability of water, the fact that its chemical quality is not always ideal (drinking water, even when suitable for human consumption, is not necessarily fit for fish), that you should not use rain water (as was once the case), and that many aquatic organisms have very precise needs, it is often essential to use demineralized or deionized water for the aquarium. This water is specially treated not only to reduce the degree of hardness but also to get rid of any harmful dissolved substances. It is advisable, however, not to use water that has simply been distilled, like that for car batteries, as it may contain toxic substances that are not removed by ordinary distillation, as well as cause other problems too.

Oxygen

Water is capable of dissolving gases as well as solids. There is an equilibrium between air and water, so that gases (oxygen, hydrogen and nitrogen) can spread freely from one to the other. Nevertheless, in comparison with air, the amount of oxygen present in water is relatively low. On average, there is 7 – 9 mg/l of dissolved oxygen, although in special circumstances this may rise to 15 – 17 mg/l or fall to 1 – 2 mg/l. But the quantity of oxygen present in water depends on many concomitant factors, including temperature. Disregarding other conditions, the lower the temperature, the greater the concentration of oxygen. In practical terms, the quantity of oxygen theoretically present is in inverse proportion to the temperature. Water may be enriched with this important gas through the effects of movement and turbulence which increase the surface of contact and exchange with the atmosphere (as happens with a fast mountain torrent). But the principal source of oxygen for aquatic habitats and for aquaria comes from the activity of photosynthesis among plants, which occurs during daylight hours. It is worth mentioning, too, that water density has an adverse influence on the quantity of oxygen present. All things being equal, that is, temperature and other conditions, a saltwater aquarium will contain less oxygen than a freshwater aquarium.

Heating and lighting

A heating system with underwater wiring fixed to the base of the aquarium ensures free flow of water in the substrate.

Most fish that are kept in an aquarium come from regions where temperature fluctuations are fairly limited through the course of the year. This means that these animals have evolved within a specific environment and adapted to temperatures that are now part of their ideal surroundings. The same can be said for fish that live in temperate cold waters, such as the Salmonidae, or even polar waters, such as the Channichthyidae, whose blood contains antifreeze substances to keep it flowing in the low temperatures found in antarctic waters.

The relationship between fish and outside temperature is even closer than this implies, however, because the body temperature of almost all fish depends on and varies with that of the water in which they live: their physiology does not permit them to keep it constant (they are "cold-blooded" or "ectothermic" animals), in contrast to mammals and birds ("warm-blooded" or "endothermic").

In northern latitudes and inside our homes it would be virtually impossible to maintain constant temperatures of 24 – 27°C (75 – 80°F), namely those needed in the aquarium to sustain tropical fish, without resorting to an outside energy source. Although in summer it may not be difficult to reach these levels, in winter the water temperature is incompatible with the survival of most tropical species. It is therefore essential to heat the water. The simplest system is to use electric heating elements, with or without thermostats. Either way, the principle is the same. The passage of an electric current through an electric resistance produces heat, transmitted in turn to the surroundings. Naturally the elements are electrically insulated and many are completely waterproof so that they can be submerged without any problems. Some function at reduced voltage, with heating wires encased in silicone which contain a resistor activated by a safety current of 24 volts. These wires are positioned on the floor of the tank, concealed and covered by the material used for lining the aquarium. A similar heating system, which will probably become increasingly popular in the future, facilitates water circulation among the sediment, due to the fact that the less dense warm water rises, while the cold water falls. The gentle flow that is created is regarded by some experts as extremely important for the plant life of the aquarium.

Naturally, the materials used for the bed should not be too compact as to impede the slow passage of water, nor too thin as to risk the plant roots being too close to the heaters and suffering damage.

The classic tubular heater is even more widely used. Plastic or glass tubes contain a resistor that is often linked to a thermostat, either bimetallic (internal) or electronic (internal or external). Still simpler models are of the fixed type, without a thermostat. The amount of heat given out is not controllable and is distributed constantly so it must be used on a sort of on-or-off basis. For this reason, such heaters must be very carefully chosen. A heater that is too small will not reach or maintain the required temperature, while one that is too powerful will raise the temperature of the water to dangerously high levels.

The estimate of necessary power is based on a few empirical rules, such as calculating 1 or 2 watts to every liter (1¾ pints) of water. Of course, the power estimate must also take into account the difference of temperature that exists between the water of the aquarium and the outside surroundings, the dispersal factor of the materials and the insulation of the tank. Using these general guidelines, it is not difficult to calculate the power needed to heat any tank, either by consulting appropriate tables or by applying the following formula with the aid of a calculator: $W = K \times C° \times V^{0.66}$, where W = power of heater in watts, K = constant equal to 0.2924, C° = increase of required temperature in relation to outside temperature, and V = volume of tank in liters. Obviously, this calculation is also necessary when regulating heaters provided with thermostats, which are always the best to use. The advantages of such heaters are self-evident. When the water has reached the desired temperature, they cut out, then turn on again when the temperature begins to drop. If we were to depict graphically the action of the two types of heater, the former might be presented by a straight line and the latter by a sinusoid.

The positioning of the heater within the tank proves to be crucial, too. In the case of a heater with a built-in thermostat, poor water circulation close to the instrument will eventually build up a mass of warmer water that will cause the thermostat to cut out. To avoid this problem, the

SIZE AND VOLUME OF TANK AND HEAT REQUIREMENT		
Size in cm	Volume in liters	Heat required in watts
30 x 20 x 25	15	50
60 x 25 x 40	60	75
80 x 40 x 35	112	100
100 x 40 x 35	140	150
100 x 40 x 50	200	200*
110 x 55 x 40	242	300*
120 x 50 x 67	402	400*
160 x 45 x 60	432	500*
200 x 60 x 70	840	750*
220 x 60 x 70	924	1,000*

* Higher temperatures can be obtained by using more than one heater.

The two drawings below show the effect of light on the aquarium. Left: with a central light the fish's shadows project towards the observer.

Right: lit from the front, the fish display their colours and iridescence.

Below bottom: heater with thermostat incorporated.
Below middle: electronic thermostat with underwater probe.

Below: modern mercury vapour or metal halide lamps hung over the aquarium are particularly useful for uncovered tanks.

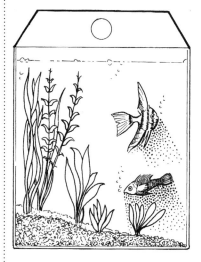

heater should be positioned close to a water current activated by a pump or an air column produced by an aerator. For similar reasons, in the case of a heater with a separate thermostat, the latter should be set up at a distance from the heater itself.

Finally, in order to guarantee optimum heating, the elements should be positioned at a slant to ensure the maximum contact between warm surface and volume, and close to the floor, but far enough away from it to allow the water to circulate. Obviously these positions are applicable only to watertight heaters, while those that are partially submerged must be secured to the walls of the tank so that they do not sink completely.

On no account should heaters be removed from the water or submerged while hot. In either instance there is a high likelihood that they will break. And naturally, remember that you are handling electrical apparatus and be sure to take the normal precautions.

LIGHTING

Light and heat are, of course, closely associated. Lighting is important in the aquarium for aesthetic reasons, but above all for the benefit of plants and fish and for its influence on several fundamental environmental conditions. One important point to remember is that natural light alone cannot be used, primarily because it is impossible to control, so you must resort to artificial illumination. The use of filament bulbs is restricted to very small aquaria, whereas fluorescent tubes can nowadays satisfy every need. For large aquaria, in particular, increasing use is being made of mercury vapour or metal halide lamps, which are hung free over the aquarium and thus do not need to be covered.

As is the case with the heating installation, the power of the lighting, too, must be carefully regulated. Here, particularly, everything depends on the plants. Light is certainly the principal factor in influencing the growth of vegetation, and for this reason you have to be aware of the requirements of the individual plants. Some like very bright lighting, others prefer it to be softer. Despite this, there are certain elementary rules to be followed, and as a rough guide you should calculate 10 watts of fluorescent lighting to every 30 cm (12 in) length of tank. This applies to aquaria with a depth of 35 – 40 cm (14 – 16 in). Another rule of thumb gives the power needed for aquaria of not more than 50 cm (20 in) in depth, which some consider the maximum for the use of fluorescent lights, as equivalent to 0.25 watts per liter (1¾ pints). Finally, it is possible to make a more precise calculation based on the lumen output of the light, this being the amount of radiant energy emitted. Dividing the lumenal value of the light (this information should be given on the packaging) by its power in watts will give the light's lumen output. Bearing in mind that the optimal figure for plant growth is 30 – 50 lumen per liter (1¾ pints), it is then simple to work out the necessary power of lighting.

Fluorescent or neon tubes are nowadays produced in a vast range of types, based on the characteristics of the light emitted: cool lights accentuate colours; daylight lights reproduce, more or less faithfully, the solar spectrum; phytostimulant lights emphasize the red–orange and blue–green fields; and black light is for special effects. As a rule, it is best to use a combination of these lamps because no single one, on its own, will guarantee perfect results. There are those, however, who recommend using only daylight lights in saltwater aquaria in order to avoid a proliferation of algae. In this regard, it is worth mentioning that for some species, for example angelfish, the presence of algae in the aquarium is considered by experts to be desirable. It is increasingly common for ornamental aquaria to be furnished with mercury vapour (HQL) or metal halide (HQI) lamps. The advantages of this form of illumination relate mainly to the quantity of light produced and the high energy yield. These lamps, which in some models are available in spot or dichroic versions, are good for creating special effects or for lighting particular points instead of the uniform lighting or lighting at levels as provided by the classic fluorescent tubes. A similar effect may be obtained with the new generation of low energy-consuming lights. Both these and the older fluorescent type must be set up above and away from the aquarium,

Fluorescent lighting is usually positioned over the aquarium or, as in this case, actually fitted into the inside of the tank's lid.

thus avoiding any possible contact with the water. This means, on the one hand, that the aquarium is not covered, as is generally the case, and, on the other, that certain rules must be followed, which may be summed up as follows: a) the light should be concentrated on the aquarium and not dazzle the observer at any point around the tank; b) a safety distance should be kept between the water and the lighting so as to prevent corrosion; c) the lights should be provided with safety glass. With this method of lighting it is possible to illuminate aquaria up to a meter in depth.

Finally, there are some useful guidelines as to where lights are best placed. A central light creates annoying shadows for the observer, which tend to conceal colours and features of the fish. A light positioned close to the rear wall of the tank creates an impression of depth. On the other hand, the aquarium will look altogether brighter if the lights are placed at the front wall. In this way, shadows will be projected backward, and the fish will appear more colourful. The use of reflectors is essential if you want to obtain a particular light arrangement. Acting like mirrors, they concentrate some of the luminous energy in the required direction and prevent its dissipation elsewhere. In rows of aquaria these reflectors are built into the lids of the tanks, but it is not difficult to make them yourself with sheets of kitchen foil.

PERCENTAGE OF EMISSION OF LIGHT WAVELENGTHS BY SOME TYPES OF STANDARD 40–WATT FLOURESCENT LAMPS FOR AQUARIA

Emission band	Cool white light %	Warm white light %	Daylight %	White Light %
Ultraviolet	1.7	1.5	2.1	1.9
Violet	7.6	5.2	9.6	6.4
Blue	21	13.1	28	16
Green	24.8	20.6	27.4	23.2
Yellow	18.4	23.5	14.5	21.1
Orange	17.9	24.3	13.2	21.1
Red	8.6	11.8	5.2	10.3

Emission band	Incandescent-flourescent %	Gro-Lux %	Natural light %
Ultraviolet	0.7	1.42	2.1
Violet	2.3	9.75	7.6
Blue	7.5	27.3	16.7
Green	20.1	14.1	20.9
Yellow	13.4	1.43	12
Orange	21.8	6.1	17.5
Red	34.2	39.9	23.2

Aeration and filtration

Columns of air bubbles constitute one of the most characteristic features of an aquarium. They oxygenate the water and set up beneficial internal currents.

AERATORS AND PUMPS

The importance of oxygen in the aquatic environment has already been underlined. As regards aeration, however, a main purpose, despite what is generally believed, is to make the aquarium environment homogeneous. Although the column of water bubbles that rises to the surface is one of the most characteristic features of an aquarium, it does not have to play any part in providing the water with oxygen. Natural systems of providing aquatic organisms with oxygen are highly effective. That the bubbles escaping from the air stone need not be entirely responsible for oxygenation of the water is proved by the fact that the first so-called "balanced" aquaria had no aerators whatsoever and that 100 cm^2 (15½ sq. in) of well illuminated plants can produce about 100 mg of oxygen in 12 hours, enough for 50 g (1¾ oz) of fish. The function of the aerator becomes far more important at night, when the plants consume rather than produce oxygen. As a result, the primary purpose of aeration is to create water movement inside the aquarium and to facilitate exchanges between water and the atmosphere, the oxygenation of the water being secondary. In any event, the smaller the bubbles, the greater this capacity becomes, because a close relationship between their surface and the volume activates the gaseous exchanges within the aquarium.

The simplest and most popular aquarium aerators make use of diaphragm pumps which, thanks to the rapid vibration of a rubber membrane, compress air into a chamber by means of a valve, this air then being blown into the aquarium through a tube connected to the air stone. The models available come in a wide range of powers and should be chosen according to tank size. The best aerators are often fitted with a power regulator so that the flow can be adjusted to actual needs. One of the drawbacks of these machines is that the internal mechanism is often rather noisy. Trying out the aerator before buying it will avoid this problem. Certainly it is important to make sure beforehand that replacement parts are freely obtainable so that the entire machine does not have to be sent back if the rubber membrane snaps after too many hours of operation. Less frequently seen are aerators worked by a separate pump, piston or turbine; these are mostly intended for large-scale aquaria.

A fundamental consideration when purchasing any aerator is how efficient and how powerful it is going to be. This should be reckoned, bearing in mind the counterpressure exercized by the water and the resistance of the tube and the air stone; an air pump working well in the shop may not do so well once placed in the aquarium at home. Another essential point is always to position the aerator above the water surface so that if it cuts out, the water does not flow along the tube as through a siphon, flooding the device. To avoid this kind of accident, you can use a one-way valve inserted along the tube between the aerator and the air stone, so as to prevent the water flowing back.

In many cases, by using the principle of air lift, an aerator can be transformed into a true aquarium system. The method, invented by Carl Loescher in 1797, is founded on the principle that air bubbles forced to flow through a tube tend, as they rise naturally upward, to take along a mass of water whose volume depends on the quantity of bubbles, the

diameter of the tube and the length of its submerged part. Actually, the pull depends not so much on the bubbles as on the fact that they form a lighter mixture of air and water that therefore tends to float. The volumes of water that it is possible to shift in this way are greater than might be believed and this explains why so-called undergravel filters or other filters, both internal and external types, function so effectively on the air-lift principle. Applying this system to the aquarium achieves the double result of oxygenating the water and creating a water flow suited to the functioning of a filter with a single apparatus.

Pumps are certainly more suitable than aerators for circulating water. The choice here is considerable. Pumps of various capacities, for fixing either in or out of the water, are commercially available, some of which can move 100 – 300 liters (22 – 66 gal) an hour. Others, as we have seen, can be attached to internal or external filters. Not only do pumps guarantee the necessary flow of water between the tank and the filters (see next section), they also prevent zones of stagnant water forming in the aquarium and create uniform conditions, setting up currents that closely resemble those exploited by the plants and fish in their natural environment. In fact, strong circulation (as would occur, for example, on barrier reefs where the ocean waves continually circulate the water) encourages the fish to behave in a natural way, stirs up and freshens the water for the plants and strengthens their leaves simply by subjecting them to continuous movement.

FILTERS

Although it is possible to set up an aquarium without filters, this limits both your choice of fish species and the total numbers. The filter, in fact, is the key to successful aquarium maintenance, and it is no exaggeration to say that the aquarium should ideally be planned around it. Its importance hinges upon the fact that an aquarium, although artificial and made to measure, remains an environment subject to natural laws, especially those governing the cycle of organic matter. Inside the aquarium there are plants (primary producers), animals (consumers) and bacteria and fungi (decomposers). In natural surroundings these three categories live in equilibrium and are closely dependent on one another. Plants have the job of producing organic matter from simple substances by photosynthesis. Organic substances become the food of herbivorous animals which in their turn are preyed upon by other animals and so on, according to the principles that

control the food chains. But every living being, animal or vegetable, produces waste matter and eventually dies. If there were no opportune mechanisms for recycling this material, life as we know it would no longer exist because many of its components would already be depleted. An infinite variety of bacteria and fungi put these substances back in circulation after decomposing and transforming them so that they can once more be assimilated by plants, thus initiating the cycle anew.

The same happens in the aquarium. Here the natural balances are not always applicable, so that after a while the accumulation of some decomposed substances threatens serious harm to the aquatic habitat.

Below bottom and middle: filters and pumps are the very essence of the aquarium.
Below: an external filter taken apart to show the various pieces and materials that compose it.
Opposite: this attractive aquarium populated by colourful fish and plants is the result of all essential equipment being in good working order.

It is the task of the filter to correct this state of affairs by recycling the water, purifying it and thus rendering it suitable for the fish in the aquarium.

Fundamentally, filters are distinguished by their action (mechanical, biological and chemical) and by their position in relation to the aquarium (internal or external).

How filters work

The simplest filter is the mechanical type. It works like a kind of sieve, trapping the waste substances present in the water. The finer the filter material, the more efficient the mechanical action is. However, the very effectiveness of the filtering medium (as, for example, diatomaceous earth) provokes the risk of rapid blockage, so that powerful pumps are necessary for the passage of water. Also for this reason mechanical filters are fitted with layers of material of increasing fineness, starting with the one that first makes contact with the water. In this way the water is gradually cleared of suspended substances, preventing them from building up in the less permeable layers and causing an obstruction. As a rule the first layers utilize nylon wool or gravel, often followed by sand of varying fineness. Best results are achieved by the special acrylic resins or foam, available in blocks and shaped to measure, which are structured rather like sponges, functioning in much the same way by retaining the evenly distributed impurities in their cavities. A useful tip in the preparation of such a filter, especially of the internal type, is to keep the various materials in separate perforated bags. Mechanical filters have to be cleaned at regular intervals and it is certainly easier if you can extract each filtering medium separately rather than extracting it bit by bit.

For satisfactory filtering action, the water should preferably flow through at a speed equivalent to half the volume of the tank every hour in the case of a freshwater aquarium, and once the volume per hour in the case of a freshwater aquarium. This provides a useful indication as to the choice of pump for operating the filter.

Similar to the mechanical filter, particularly in respect to the granular structure of the medium, is the so-called chemical filter, the name of which is misleading. Although this system can use ion-exchange resins or particular zeolites to get rid of some of the waste substances (especially nitrogenous compounds), it is as yet too expensive for domestic fish keepers. However, ion-exchange resins are commonly used to lessen the hardness of the water. At present, chemical filtration usually involves the use of activated carbon. This is a particular type of carbon with granules of specific and, as far as possible, uniform dimensions. The basic feature of this product, due to its preparation, is that its surface is extremely porous, so much so that a gram of activated carbon may have a surface of between 500 and 1,500 square meters (595 and 1,785 sq. yards). This allows the carbon to adsorb, i.e. accumulate in its pores, even substances dissolved in the water, such as those that give it colour. But activated carbon does not discriminate between useful and harmful substances, and this includes medicinal elements used in the treatment of disease. Consequently, filters with activated

carbon cannot be utilized in tanks designed to cure diseased fish or when therapeutic treatment is to be carried out; an additional reason is that certain medicines, such as phenoxyethanol, may substitute themselves in the pores for toxic substances already adsorbed, thereby releasing the latter, rapidly and in large quantities, with devastating effects on the plant and fish populations. Finally, it is worth noting that activated carbon tends gradually to lose its adsorptive capacity due to the progressive inactivation of the pores, transforming itself into an ordinary substratum for the biological or mechanical filter.

Sooner or later mechanical or chemical filters are transformed naturally into biological filters, which work thanks to the action of bacteria carrying out the processes of decomposition throughout the aquarium. It is important to locate them in areas where oxygen is always abundant. The fundamental process that occurs in these highly effective bacterial filters is the breaking down of nitrogenous compounds. The intermediate stages of these processes comprise the transformation of ammonia (NH_3), one of the principal products of fish waste, and of ammonium ions (NH_4^+) into nitrites (NO_2) and then nitrates (NO_3). The oxygen content (O) in these compounds increases progressively, and this explains why it is indispensable if these reactions are to be carried out properly, particularly since both ammonia and nitrite, even in low concentrations, are toxic compounds for fish.

A biological filter, by its very nature, needs some time for running-in so that suitable and sufficient bacterial populations, which are in short supply in a newly built aquarium, may develop. This growth is in direct proportion to the decomposable organic matter and to the temperature. Thus cold aquaria need more time than tropical aquaria to "mature." For this reason fish are introduced gradually because the purifying capacities of the filter are initially very low, recycling the waste products of only a few fish. On average it takes three weeks for a tropical aquarium to mature, and about twice as long for a temperate one. The running-in period may be reduced by using preparations of freeze-dried bacteria and specific enzymes, nowadays available to aquarists following their successful use in industrial purification plants, which work on more or less the same principles as aquarium filters.

A special type of biological filter uses algae and aquatic plants as well as bacteria for the removal of waste products. In marine aquaria, where it is difficult to introduce plant populations comparable to those used in freshwater tanks, it is possible to utilize shallow algae filters which are nevertheless quite large and therefore more appropriate, because of the space they occupy, for big aquaria. For freshwater aquaria it is feasible to set up tanks, suitably filled with plants, alongside those containing the fish. In both cases filter tanks have to be well lit in order to stimulate a luxuriant growth of vegetation. The water is filtered from the plants which extract from it carbon dioxide and inorganic compounds, releasing oxygen and organic compounds. The system is useful when breeding herbivorous fish or others which like to supplement their diet with vegetable food.

The materials available for use inside an aquarium and for preparing a filter differ according to the type of tank for which they are intended. However, all of them, quite apart from their purpose, must be guaranteed nontoxic and cleaned prior to being introduced.

MATERIALS FOR THE AQUARIUM

Material	Mechanical properties	Biological properties	Adsorbent properties	Sea water	Fresh water
Nylon wool	●	●		●	●
Basalt pellets		●		●	●
Anthracite	●	●		●	●
Quarz	●	●		●	●
Activated carbon	●		●	●	●
Activated fibers	●		●	●	●
Coral sand	●	●		●	
Crushed shell		●		●	
Lavic rock		●		●	●
Noncalcareous gravel		●			●
Calcareous gravel		●		●	
Peat	●				●
Ion exchange resins					

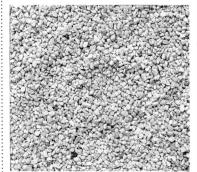

TYPES OF FILTERS

External filters

These filters can be used for aquaria that are intended to be completely filled with fish and plants. As the name suggests, they are fitted outside the tank and consequently they are, or should be, easily accessible. As a rule they are installed near the upper edge of the aquarium, and linked to it by tubes. In theory, the water can be made to flow in by a pump and to exit by an overflow, or it can be withdrawn and returned to the tank by a pump situated in the filter, which fills up through a siphon. Other pumps, of course, work on the air-lift system already described. Of the two, the siphon system seems preferable because it cuts out when the water in the filter reaches the same level as the water in the tank. In the other case there is always a risk, however remote, of the exit hole being blocked for some reason, causing the water to overflow. Other external filters consist of airtight containers furnished with centrifugal pumps. The water penetrates to the bottom of the container and after passing through the filtering materials is sucked up by the pump and returned to the aquarium. These filters can also be positioned below and some distance away from the tank because, obviously, the pump has the necessary power to overcome the pressure of the column of water above.

Internal filters

These filters, as a rule, have a bigger volume than external filters, occupying up to 15 – 20 per cent of the tank and thus reducing the space available to plants and fish. Like the external type, they are divided into compartments containing the filtering medium, and the water follows a path through these. In most cases the water to be filtered flows from the bottom, where the largest amount of waste generally accumulates, and through a series of holes protected by grating so that the fish cannot get in. From there the water rises and, on the principle of communicating vessels, flows into other compartments where it is thoroughly filtered. Inside the filter is a submerged pump of suitable power, responsible for constantly changing the water. A good rule is to place an air stone connected to the aerator into the filter as well, where the oxygen consumption is high. The maintenance of internal filters is more complicated than that of external types, particularly if the tank is partially or wholly enclosed. In fact, when the filtering materials are removed, the accumulated substances may sink back into the tank. For this reason, as previously suggested, these materials should be kept in bags so that they can be quickly extracted without problems. Small internal filters, similar to external ones, are also available with a centrifugal pump and insert, mechanically operated.

One particular type of internal filter is the so-called undergravel filter. With this system the whole base of the tank is converted into a large filter. It should be installed at the same time as the tank is being filled, for if this decision is made later it will entail partial or total dismantling of the tank and starting again from scratch. The principle upon which this system is based is that the floor of the aquarium is itself an ideal mechanical and biological filter, provided that the water in the tank flows evenly and regularly throughout. To achieve this, special filter plates, which can be individually assembled to cover the available surface, have to be placed beneath the aquarium gravel. One or more of these plates are furnished with a hole for a tube. The tube is connected to a pump or to an air stone, which sucks up the water from this double base layer and creates the necessary direct current from top to bottom. In this way the water passes through the base materials, is sucked up in the tubes and then returned, cleaned and filtered, to the tank. However, the flow must not be so swift that it may prevent adequate treatment of the water.

This type of filter will in time gradually lose its filtration capacity and be damaged by medicinal treatments that reduce the efficacy of the bacterial populations. Nor is the use of such filters compatible with the lifestyle of certain markedly benthic fish species that burrow on the bottom, because these will eventually alter the uniformity of water circulation in the filter, an essential condition of its functioning properly. Finally, it goes without saying that replacing this type of filter entails complete emptying of the aquarium. This is not too difficult a task with small aquaria but likely to pose serious problems with larger ones.

Equipment and accessories

Control of the main chemical and physical properties of the water need present no difficulties. For example, the pH can be measured by means of colorimetric tests or special instruments.

In the preceding sections we have considered the various components of an aquarium and the necessary measures to guarantee the wellbeing of its animals and plants, based on their principal needs: oxygen, light, heat and elimination of toxic substances. In addition to these components are all those accessories that are more or less indispensable to the proper functioning and maintenance of the aquarium. Basically, they compromise: a kit for chemical analyses, cleaning equipment, timers, automatic food distributors, ozone reactors, protein skimmers, ultraviolet sterilizers and dispensers of carbon dioxide.

CONTROL OF WATER QUALITY

As mentioned in the section on water and its characteristics, this compound must possess precise requisites (acidity, hardness, absence of pollutants) to satisfy the vital needs of fish and plants. Chemical analysis systems are therefore essential for regulating the water quality and should be part and parcel of the maintenance equipment. In time you will certainly realize that the inhabitants of your aquarium are themselves the best indicators of how healthy the tank is, but it is useful, too, to know how to verify your suspicions by means of some simple chemical tests.
For normal aquarium use there are now kits which combine extreme simplicity and precision in order to provide all the necessary information on the subject. All the available tests are based upon the method of colorimetry (colour comparison). The addition of certain substances, varying from one analysis to another, brings about a change in the colour of the sample of tank water, and the significance of this is interpreted by means of a comparative scale. The simpler models consist of a plastic card bearing a range of colours which are compared to the colour obtained during the analysis. Other models have colorimetric discs, which are more precise and permit more detailed analysis. In any event, it is important to work with a good source of light (a room well illuminated by the sun is ideal) in order to evaluate the colours obtained with accuracy. This applies mainly to the lower values of the scale.

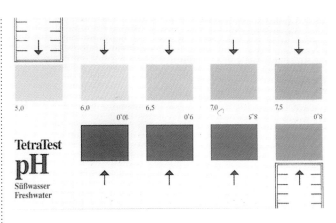

THE PRINCIPAL ANALYSES

The list of possible analyses becomes longer year by year, but the following are those that may be regarded as essential.

Water acidity and alkalinity (pH). This test enables you to establish by how much the water in the aquarium has shifted from neutrality (a value fixed at pH 7) and assess if this shift still provides ideal conditions for the

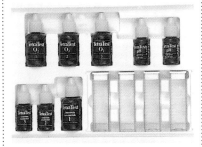

fish and plants. Some organisms, in fact, prefer water which is slightly acid (values below pH 7) while others, notably marine species, live permanently in alkaline water (values above pH 7). The appropriate tests measure the pH value of fresh water on a scale ranging from 4.8 to 7.5, and of sea water on a scale of 7.5 to 9.
In addition, there are electronic instruments, portable and not too expensive, which give a quicker and more accurate analysis simply by dipping the probe into the water of the tank. However, care must be taken to keep these instruments in proper working order, and they require calibration with liquids of known pH prior to use.

Hardness (GH and KH). As already mentioned, this parameter indicates the quantity of dissolved salts in the water, which may influence the inhabitants. Some species, in fact, prefer fairly soft water whereas others do not seem to be greatly affected either way. The tests available measure both the temporary or carbonate hardness (KH) and general or total hardness (GH) of the water. It is worth mentioning in this context that in freshwater aquaria, because of the water's chemical constituents, it is inadvisable to introduce calcareous material which dissolves slowly and thus keeps the hardness level high, to the severe detriment of the whole aquarium.

Ammonia, nitrites and nitrates (NH_3, NO_2, NO_3). The importance of these nitrogenous compounds has already been described in the section on filters. The tests most frequently used in seeking and identifying these compounds are those for the identification of nitrites, which can be toxic and dangerous to animals, and whose presence is the most important indication that the filter system is malfunctioning.

Oxygen (O_2). This is a more laborious analysis than the others, only usually carried out by the more demanding aquarists, given that this variable is seldom too low in a well maintained aquarium.

Carbon dioxide (CO_2). This test is especially useful for aquaria containing many plants since this gas, usually in short supply, plays a significant part in plant metabolism.

Chlorine (Cl_2). Very often the water used for setting up an aquarium comes straight from the tap, and because drinking water is treated with chlorine as a disinfectant, it may be useful to measure the concentration of this substance. However, prolonged and energetic aeration is normally sufficient to rid the water of chlorine, rendering this analysis unnecessary.

Specific gravity (SG). This applies only to salt water and the test helps to regulate salinity (related to SG), otherwise very complicated to measure (*see* table). Specific gravity is measured with simple instruments called hydrometers, which resemble floating thermometers or are in the form of transparent plastic boxes with a needle. The first type can be kept permanently in the aquarium and is suitable for water of any given density. The graduated scale makes it possible for the salt content of the tank to be checked day by day.

A decision has to be made as to which of the analyses listed above are essential. It is vital to include the test for nitrogenous compounds, which are related to the efficient working of the filter and which provide an indication of the overall health of the aquarium and the possible need to take appropriate measures (water change, replacing filter medium, etc.). Checking the pH

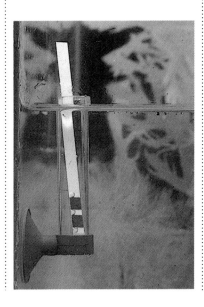

is also important, given its influence on the life of the organisms and the toxicity of ammonia. These analyses are to be recommended both for the marine and freshwater aquarium. In the latter case it is advisable also to test the water for hardness, while in the former it is important to test the specific gravity which, as already explained, can be measured continuously without intervention. As for the other tests listed, this is a matter for individual choice, seeing that shops specializing in test kits often sell colorimetric tests for a varied range of analyses (copper, cyanide, iron, etc.).

CLEANING ACCESSORIES

Although these devices are very simple, they are extremely useful for prolonging the life of the filter and keeping the aquarium clean. Particularly easy to use are siphons which may be mechanical (electric or worked by the aerator) or manual (mouth suction), the latter consisting of simple glass or plastic tubes with bubbles, or of more hard-wearing rubber tubing, similar to what is used for decanting wine, with a convenient closure valve. These accessories suck out the detritus, which inevitably accumulates on the floor of the tank without finding its way to the filter either through absence of currents or because it is too heavy. Cleaning the bed every three to four weeks and topping up with clean water (after checking the quality) will enable the filter to last longer.

Another cleaning accessory that will certainly be needed is a scraper, used for getting rid of the ugly coating of algae that is bound, sooner or later, to form on the walls of the aquarium, obscuring a clear view. Some types consist of a magnetic sponge placed inside the tank which is operated by the magnet on the outside. As will be obvious, the device must never be moved rapidly (the sponge will come loose, lose contact with the magnet and fall to the bottom of the aquarium, from which it will have to be fished out); moreover, transferring it from one wall to another is difficult because of the tank corners. A good solution is to attach a float to the internal sponge with a piece of thread, in such a way that it can just be grasped to recover the sponge. Other scrapers look like ordinary window-cleaning brushes, with hinged heads and blades for scraping the layer of dirt off the glass. These must on no account be used for plastic tanks and need a good deal of patience because of the limited blade size and the constant adjustment for cleaning.

You will also need equipment for plant care. As a rule, it is not ideal to have to plunge your arm into the tank to get rid of dead leaves or replace a plant that the fish have dislodged. For this purpose you can get long forceps, tongs and planting sticks which will enable you to do all

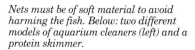

these odd jobs with little trouble and great precision, and see everything you are doing through the glass.

EQUIPMENT FOR WATER QUALITY CONTROL

Automatic devices. Modern technology extends to the aquarium and there are many labour-saving automatic devices now available to perform the most complicated tasks. Time switches for turning lamps on and off, often in sequence to avoid lighting stress, automatic food distributors (although there are simpler spring-operated models), supplementary aerators, ozone reactors, etc. Timers are especially valuable when owners go on holiday and the aquarium may be abandoned for several weeks on end. It goes without saying, however, that for long-term use, such devices must be 100 per cent reliable.

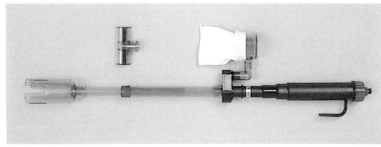

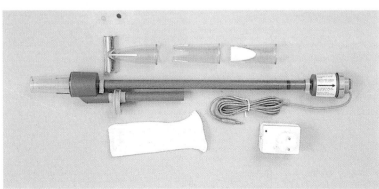

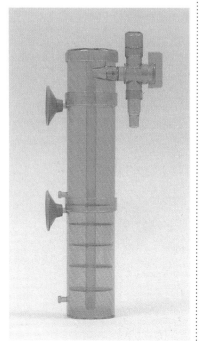

Nets must be of soft material to avoid harming the fish. Below: two different models of aquarium cleaners (left) and a protein skimmer.

Protein skimmers, ozone reactors and ultraviolet sterilizers. These are items of equipment that are not really a necessary part of standard aquarium maintenance because of their specialized functions and sometimes their high cost. The simplest of these, and probably the most useful one, is the protein skimmer. This is a simple plastic tube almost wholly submerged in the water, surmounted by a removable beaker. Inside the tube, air is drawn up through a piece of air stone to create a column of bubbles. The liquid bubbles up the tube and leaves a proteinous deposit in the beaker above the water surface. At intervals this brownish fluid has to be emptied out. The device is particularly effective and worthwhile for marine aquaria, but also useful for freshwater aquaria. What it does, in fact, is to get rid of the so-called albumen, particular proteins derived from the decomposition of the aquarium's organic substances, which would otherwise be broken down in the filter.

Sometimes the skimmer is attached to an ozone reactor, which is itself linked to an aerator. The ozone reactor gives out a succession of electric discharges (with an effect similar to that of a lightning bolt through the atmosphere) and thereby creates molecules of ozone, an unstable gas which is actually a form of oxygen with three rather than two atoms. The strong oxidizing power of this gas helps prevent many diseases in the aquarium, oxidizes the organic substances and keeps the water clear. Particularly in large tropical aquaria that contain invertebrates such as anemones, madrepores and soft corals, the use of an ozone reactor at regular intervals is very useful. Its effectiveness is diminished by increases of temperature and of pH, and it is better if the flow of ozone occurs in an opposite direction to that of the water. In any case, before purchasing this apparatus, you must be fully informed as to its exact method of use, its associated risks and the materials to be employed (silicone tubes only, because ozone destroys many kinds of plastic). Ultraviolet (UV) sterilizers are a much simpler proposition, useful (though not essential) for curing infections among the fish. These are suitably insulated ultraviolet lamps, designed to sterilize the water pumped around by killing the bacteria and protozoa. The lamps, if used continuously, will last a few months and must then be replaced as they gradually become ineffective.

Scissors and planting sticks are useful for tending the plants without too much disturbance of the fish.

CO_2 dispensers. A final addition to some or all of the accessories mentioned above are CO_2 dispensers, particularly valuable in aquaria filled with plants. They are small cylinders containing carbon dioxide, linked to special dispenser tubes placed inside the tank, which permit the gas to be released gradually into the water. The use of this apparatus entails a fairly frequent check on the pH content of the tank, for this will tend to diminish as the CO_2 dissolves in the water to form carbonic acid (HCO_3).

Preparing the aquarium

The proper setting up of an aquarium, such as the one opposite, depends largely on a detailed plan (below) and a careful choice of occupants.

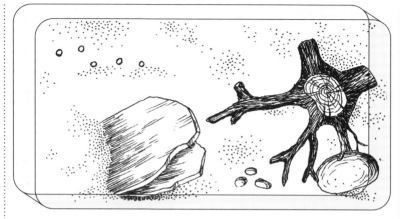

WHERE TO SITE THE TANK

The first step in the preparation of an aquarium is to decide where it is to stand. As already mentioned, this is an important decision, in which you must carefully weigh up all the pros and cons of a position. The aquarium should be in full view, but all wires, tubes and similar contraptions need to be hidden. At the same time the tank must be easy to maintain. It should not be placed near cold air currents or too close to windows and must be well away from sources of vibration and fumes. Once the ideal position is chosen, you can set up the tank, making sure to insert a layer of extruded polystyrene at least 1 cm (½ in) thick between the tank and its support. This will protect the base from possible breakage and will provide thermal insulation. Be sure, too, to wash and rinse the tank thoroughly beforehand so as to get rid of any bits of material used in constructing it or perhaps left behind in the course of transporting it.

DESIGNING THE AQUARIUM

At this point it may be helpful to draw up a rough plan of the finished aquarium (*see* figure above). This will give you a clearer idea of where you are going to situate the aeration tubes, the air stones, the heaters and all the other accessories you are likely to need. Most of these items need to be as unobtrusive as possible, so arrange your plan around their positioning. First, fix the tubes in place, then you can proceed with the preparation of the gravel bed or, if the tank includes this, the fitment of the undergravel filter. The material for the aquarium bed must be carefully chosen according to the kind of aquarium desired. The substrate is an important constituent of any aquarium, not only visually but also for the wellbeing of the fish and plants. The choice has to be made bearing in mind such factors as form, grain size, sequence of layers, colour and (perhaps most important of all) chemical composition. For example, coral sand is ideal for a tropical aquarium, helping to keep the pH value constant, but absolutely unsuitable for a freshwater tank, for which sand or siliceous gravel must be used. Equally prohibited in such aquaria, because they are not only out of place but potentially harmful, are branches of madrepore coral and seashells. As regards colour, it is best to choose a dark one that imitates the tones of natural beds which, with the exception of coral sand beds, tend not to be very bright. Moreover, a dark background will provide an effective contrast to the colours of the fish, particularly the more brilliant ones, and the green of the plants. You can, if you wish, soften the overall impression of darkness by interspersing groups of lighter and variously coloured rocks.

When it comes to choice of substrate (on average 2 cm^3 to every 100 cm^2 of surface), bear in mind that it must provide room for the plant roots to develop sufficiently to take firm anchorage and also, if you are using an undergravel filter, to ensure that there is adequate filtration. As a general rule, the finer the substrate the less thickly it should be laid so as to avoid the possible formation of pockets without oxygen. So it is advisable to use grains of 2 – 6 mm and for the layer to be 5 – 10 cm (2 – 4 in) thick, following the rules indicated below.

Where you are not installing an undergravel filter, you can create so-called terraces to vary the nature of the bed and set up raised features with are just heaped-up piles of thick substrate. Such terraces can be made of fiberglass, with plates of nontoxic plastic or clay, or even polystyrene blocks. The last, because it is light and easy to work, is ideal for making artificial beds with hollows and caves, without resorting to the more laborious fiberglass. The polystyrene floor can be smeared with silicone glue and covered with sand or gravel to conceal the white colour, or it can be painted, provided, of course, you use nontoxic, waterproof colours. Rocks, terraces and artificial floors are also a good way of hiding tubes and filters, apart from giving the tank a more natural look.

Wood and cork, too, are excellent materials for tank design, and even more natural in appearance. With luck, you may be able to pick up bits of wet or partially petrified wood beside an unpolluted stream or river; failing this, you can buy them in aquarists' shops.

If you wish, the aquascape can be further enhanced by the addition of scenic friezes which can be bought in

lengths and set up either inside or outside the aquarium to give a better impression of depth and space. A greater feeling of expanse can also be achieved by gradually raising the level of the substrate towards the rear wall (1 cm [½ in] to every 8 – 10 cm [4 in] of wall height), thus reducing the quantity of refracted light. Equally, you can arrange rocks to create fissures and cavities in order to increase the sense of distance. Lighting, too, can be used to obtain effects that give the tank a sense of additional space.

FOCAL POINTS

When planning the aquarium and deciding on focal points, it is helpful to bear in mind certain basic principles of perspective, rather as if you were designing a stage set. There are four such points in the aquarium, although you can only use two, situated on the same diagonal. Place the tall objects here to create effects of depth. It is a good idea to make a scale drawing of your projected aquarium beforehand to get the measurements and focal points right. Remember, too, to make a harmonious scene, making each feature part of it so that the finished design does not look fragmented and confused. Avoid, also, inaccessible corners that might impede the circulation of water and add to the difficulties of cleaning.

FILLING THE AQUARIUM

Once the arrangement of the tank is complete, you can fill it with water. To avoid any risk of disturbing the substrate once it has been carefully laid, protect the bed with a piece of plastic. Then, hold the hose over the rocks or terraces or pour the water slowly and gently over a bowl or tub,

Furnishing an aquarium involves following certain steps. Having checked the strength of the tank and chosen its final position, you can begin to introduce materials for the bed (1) after thoroughly washing them. Next, set the pumps and tubes in place (2). When positioning the filter materials (3), take into account the entry and exit points of the water. The next step is to arrange the decorative items (4) according to your plan. Filling the tank (5) should be done gradually so as not to stir up the substrate.

FISH AND THEIR HABITATS

Fish	Area	Fish	Area	Fish	Area
Abudefduf troschelli	2, 3	*Dascyllus tzimaculatus*	1	*Acanthophthalmus semicinctus*	3, 5
Acanthurus achilles	1, 2, 3	*Diodon holacanthus*	2, 4	*Alestes longipinnis*	1, 3, 4
Acanthurus lineatus	1, 2, 3	*Dunkerocampus dactyliophorus*	3, 4	*Anableps anableps*	3, 5
Acanthurus triostegus	1, 2, 3	*Equetus lanceolatus*	1	*Betta splendens*	5
Acanthurus leucosternon	1, 2, 3	*Gobiosoma oceanops*	1, 2, 4	*Callichthys callichthys*	5
Acanthurus lineatus	2	*Gomphosus varius*	1, 2, 3, 4	*Carnegiella strigata*	1
Aeoliscus strigatus	4	*Gramma loreto*	1, 2	*Corydoras aeneus*	2, 5
Amphiprion ocellaris	2, 4	*Heniochus acuminatus*	1, 4	*Cottus gobio*	3, 4
Apogon maculatus	2, 3	*Histrio histrio*	2	*Gasterosteus aculeatus*	3
Arothron reticularis	1, 2, 4	*Holacanthus bermudensis*	2, 3	*Geophagus jurupari*	3
Aspidontus taeniatus	1	*Holacanthus rufus*	1, 2	*Gnathonemus petersi*	2, 3, 4
Balistapus undulatus	1, 4	*Hyploplectrus unicolor*	1, 2	*Hypostomus plecostomus*	2, 4
Balistes vetula	2, 3, 4	*Naso lituratus*	2	*Labeo erythrurus*	3, 4, 5
Bothus mancus	4	*Naso unicornis*	2	*Lepidosiren paradoxa*	5
Centropyge argi	1	*Melichthys niger*	1, 3	*Lepomis gibbosus*	3, 4, 5
Centropyge flavissimus	3, 4	*Myripristis murdjan*	1, 4	*Loricaria filamentosa*	2, 4
Centropyge loriculus	1	*Labroides dimidiatus*	1, 3, 4	*Malapterurus electricus*	3
Centropyge potteri	1	*Liopropoma rubre*	1	*Pantodon bucholzi*	5
Cephalopholis miniatus	1, 2	*Opistognathus aurifrons*	2	*Perca fluviatilis*	3
Chaetodon aculeatus	1	*Pomacanthus imperator*	1	*Phenacogrammus interruptus*	4, 5
Chaetodon auriga	1, 3, 4	*Pomacanthus paru*	2	*Polypterus bichir*	3, 4, 5
Chaetodon ephippium	1	*Rhinecanthus rectangulus*	2, 3	*Protopterus dolloi*	4
Chaetodon kleini	1	*Scarus ghobban*	1, 3, 4	*Pseudotropheus zebra*	3
Chaetodon lunula	1	*Serranus tabacarius*	1, 2	*Rasbora heteromorpha*	1, 4
Chaetodon ocellatus	2, 3	*Zebrasoma flavescens*	1, 2	*Sarotherodon mariae*	4
Chaetodon ornatissimus	1	*Zebrasoma veliferum*	2	*Serrasalmus nattereri*	1, 3, 4
Chaetodon ulietensis	1, 4			*Synodontis nigriventris*	3, 4
Chromis caerulea	2, 3			*Trichogaster trichopterus*	5
Chromis scotti	1				
Cirrihitichthys oxycephalus	1, 3				
Chromis cyanea	2				
Coris gaimardi	1, 2, 4				
Coryphopterus personatus	2, 3				
Cromileptis altivelis	1, 4				
Dascyllus aruanus	1				

Although all the above tropical sea species live in the vicinity of coral reefs, they each predominate in different areas:
1 = deep outer reefs
2 = breaker zones
3 = rock pools
4 = inner lagoons.

Freshwater species are found in an equally varied range of habitats:
1 = pelagic zone
2 = benthic zone
3 = littoral zone
4 = rivers and canals
5 = stagnant or plant-rich water.

When planting an aquarium, bear in mind the plants' comparative heights, colours and growth rates before arranging them.

so the jet of water makes as little disturbance as possible. As the tank gradually fills up, everything suddenly comes to life. It is an exciting moment as the tank takes on the true appearance of an aquarium.

Once the tank is full, you can attend to the electrical fitments and test the pumps, aerators and lights. At this point it is best not to fill the filter completely. Even if the material you intend to use has been thoroughly cleaned beforehand, particles of dirt and detritus will still be present and be likely to damage the filter. Just put in a layer of filter wool and wait. The water will slowly begin to circulate, the particles will fall to the bottom and many of them will end up trapped in the filter so that eventually the water becomes clear. This is particularly useful in saltwater aquaria because it facilitates the cleaning of sediment and materials without wasting useful salt. At this point you can remove the used water and replace with new water after completing the filtering process.

Although this preparation may seem a bit tedious and complicated, it is a

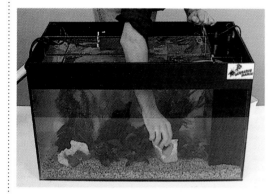

After filling the tank almost completely, you can attend to the final details. Carefully set the plants in place (top three phases). Then, after leaving the planted aquarium in working order for a few days, introduce the fish, a few at a time, making sure they do not suffer temperature shocks.

good way of ensuring that everything is working properly and to some extent it will give you time for second thoughts about any badly placed objects. Now, as you fill the aquarium for the second time, having made quite sure that the chemical and physical properties of the water are appropriate, you are almost ready to introduce some of the live inhabitants.

PLANTING THE AQUARIUM

Leave a few days to make sure that all is functioning properly. When you have satisfied yourself that the lighting and heating are operating well, and that the pumps and aerators are keeping the water circulating and well provided with oxygen, you can proceed at last to the final stages: the introduction of some plants and fish.

There are good reasons why plants should go in first. In the course of natural events, plants appeared before animals, and in the context of the aquarium, the positioning and arranging of plants entails a certain amount of time and labour. Each plant has its individual requirements and has to be considered in relation to others, forming a harmonious pattern, and they must have time to acclimate to their surroundings. If fish were already present, they would be continuously and unnecessarily disturbed while planting took place. To create a sense of movement, set your smaller plants near the front of the tank and the bigger ones towards the back and in the corners so as to cover, though not conceal completely, the glass walls. Each plant should be assessed for its needs, its growth potential and its systems of reproduction. If you understand and foresee how a plant is likely to develop, there will be no need to change its position when it is fully grown. It is best not to buy too many different species or any that are hard to blend in with others and would look out of place in the general environment of the tank. Do not forget, either, unless you have planned a plant-filled aquarium, that putting in too many plants may leave insufficient room for the fish and hide them from view. As regards colour schemes, it is desirable that plants and fish provide a contrast but it is essential, too, that they should complement one another, as they do in nature.

Place the plants carefully in position by hand rather than with a planting stick, which may damage them, especially if you have not used such a tool before and are unsure how much pressure to exert. The plant should be grasped just above the roots between thumb and index finger and placed in a hole sufficiently large for them to be spread wide. When in place, cover uniformly with soil, taking care not to bury the collar. Not all plants have roots of the same size, and bigger ones obviously have to be given more depth and width. Moreover, the roots should be cut back slightly so that only the healthy parts are retained, both to make planting easier and to help the plant to take root.

When all the plants are in, they should be given a few days to acclimate, providing them with continuous lighting for the first 48 hours.

Along with the plants, you should introduce useful bacteria and organic substances, which will help them to grow and the filter to mature. The aquarium is naturally deficient in these substances, so it is a good idea to make use of the products freely available today (bioconditioners and specific fertilizers) in order to speed up the processes both of filter maturation and plant growth. If the aquarium plan allows for it, you can introduce a few fish at this stage as long as they are the small and less demanding ones, and on no account too many. In this way you set in motion the basic cycle of organic material that transforms the aquarium into a miniature ecosystem.

INTRODUCING THE FISH

When the plants are in, settled and thriving, with all systems functioning satisfactorily, you can proceed with the final phase – introducing your selection of fish to the aquarium. Your choice of species will have depended on a range of factors that have been mentioned already in the previous sections. So we need only consider here a few practical guidelines as to the number of fish to be accommodated, how to choose them and how to introduce them. A precise calculation as to the number of fish that could be accommodated in one particular tank would need a team of experts and a laboratory equipped for the appropriate analyses. Many things need to be borne in mind: the amount of oxygen present in the tank in relation to its temperature, the consumption of this gas by the filter, the bacteria and the fish, the quantity of excrement produced, and so on. For this reason, it is generally enough to resort to certain rules, which although somewhat empirical, serve as pretty good indicators.

A shoal of Rasbora hetermorpha *perfectly acclimated in an aquarium which will be its home for a long time.*

Counting the number of fish can be done on the basis of their length, calculating that, on average, a fish measuring 1 cm (⅓ in) in length requires 2 liters (3½ pints) of water. Consequently, one fish of 10 cm (4 in) or ten fish each of 1 cm (⅓ in) will need 20 liters (4¼ gal) of water. These indications, valid for tropical freshwater aquaria, have to be modified downward for marine and for temperate freshwater aquaria, where the number of fish per liter (1¾ pints) of water should be reduced by about 30 per cent; alternatively the quantity of water per fish should be increased to 3 liters (5¼ pints).

When you calculate the volume of water contained in the tank, remember to subtract from the cubic measurement of the tank the volume occupied by sand, gravel, rocks, filter and other decorative elements. Another rule of thumb is that every 1 cm (⅓ in) of fish (excluding the tail) should correspond to 75 cm² (11½ sq. in) of surface area. Here too the figure applies only to tropical freshwater aquaria. For temperate freshwater and marine aquaria the appropriate areas must be increased to 180 cm² (28 sq. in) and 300 cm² (46½ sq. in) respectively.

Now that you have some idea as to how many fish to accommodate in your aquarium, all that remains to do is go out and buy them. Select a few at a time, relying upon your experience, instinct and taste, plus some expert advice, should you need it, on the compatibility of different species and the dimensions to which they are likely to grow. For example, ten fish of 3 cm (1 in) that are each capable eventually of reaching 10 cm (4 in) will have problems in living together later as the aquarium becomes overcrowded.

When you buy, try to follow these simple rules:
1) Only purchase fish that have been kept for an appropriate period in quarantine.
2) Make sure that the colour is typical of the species and variety required. Any variations that are not due to special situations (for example, breeding period) may in fact turn out to be signs of disease.
3) Take a good look at the behaviour and form of the fish: difficulty in swimming, sunken belly, emaciated appearance or frayed fins will help to identify specimens that are not entirely healthy.
4) Make certain that other fish in the same tank are healthy. The presence of dead fish on the bottom should arouse immediate suspicion.

Once you are at home, before introducing the fish to their new surroundings, it is a good idea to switch off the aquarium lights and go over for the time being to ordinary diffused room lighting. Before tipping the fish out of their plastic bags, dip the whole bag into the water to test the temperature. Then pierce a few holes in the bag so that the water in it can gradually mix with that of the aquarium. Only then should you open the bag completely to let the fish out and swim free. Continue to keep the lights dim until the fish get accustomed to the new tank and appear to be swimming around confidently. After a few hours turn on the lights, give the fish a little food and settle down to watch them, checking they seem happy and show no signs of suffering.

Aquarium maintenance

Owning an aquarium means adhering to certain basic rules and knowing exactly how to take care of its fish, plants and invertebrates.

Regular maintenance of an aquarium is absolutely essential because otherwise any problems, generally due to serious changes of condition, will become too difficult to resolve. Even in the most modern, efficient aquarium, there are still jobs that have to be done regularly by you. The following are the most important tasks.

DAILY

Allow ten minutes or more each day for just looking at your aquarium. This is as useful as it is enjoyable. It will enable you to check that the temperature is right, that the various systems are working properly, and that the fish are behaving and feeding normally. As for food, do not assume that tossing a few scraps now and then into the tank is enough. Supply a little at a time and try to make sure that all the fish, not only the biggest and strongest, get enough. Between one feeding session and the next, wait until the fish have used up all the food previously given. In this way you will avoid unused organic substances piling up on the floor.
Get to know your fish, too, and count them regularly to make sure that none of them is dead or hidden from view while decomposing. Dead fish should be removed as quickly as possible so that the water does not become contaminated. Get rid, too, of dead leaves and keep a watch on the condition of the plants. They also can provide valuable information as to the general state of the tank.

WEEKLY

After switching off the electrical contacts, siphon the bottom of the tank to eliminate dirt, which is bound to accumulate in the dead ground that exists in all aquaria. Check the plants and the efficiency of the air stones and pumps, of the air and water feed pipes, and clean them if they seem dirty or clogged. Save, if you think it opportune, the water extracted during the siphoning, after filtration to eliminate the dirt.
Check (every two weeks will do if you think everything is in order) the chemical values of the water by means of the appropriate tests (*see* section on Equipment and Accessories) and do what is necessary to restore those values that appear abnormal.
If need be, clean away the algae on the walls, using a scraper.

MONTHLY

Renew the water level to compensate for evaporation, ensuring that the characteristics of the added water are compatible with those of the aquarium. Remember that for marine aquaria there is no need to top off with salt water but only with fresh. The salts, in fact, do not evaporate, only the water.
If your chemical examinations reveal an excessive build-up of nitrogenous compounds, the time has come to replace some of the water, changing about 20 per cent of it. Pay attention to the quality of water used for the change and add it slowly.
Get rid of the dirt that has accumulated in the first compartment of the filter and give a rough clean to the mechanical filter. Do not touch its biological part.
Check the contacts of all the electrical apparatus, especially those systems inside the tank, to prevent oxidization and loss of efficiency. Use vinegar to eliminate calcareous crusts, or water to wipe off the salt deposits that may have formed on lamps.

ANNUALLY

Check all the plants, pruning those that have grown too much and replacing dead ones if necessary. Replace fluorescent tubes, which gradually wear out.
Check the aerators and the state of their diaphragms.
Clean the pumps and carry out any maintenance as recommended by the manufacturers. Replace the mechanical filter materials and, if need be, some of the biological filter as well (especially if the nitrate values remain very high and the nitrites will not decrease as rapidly as before). In the case of the biological filter, replace only those materials that have really become too dirty, and merely clean the others with a jet of water, but not so as to reduce too much of the existing populations of helpful bacteria. Remember that if cleaning is too drastic, it may mean you have to start off the aquarium again virtually from scratch, although the new filter will mature more quickly because the bacterial population is already abundant and balanced.
Check all thermostats and electrical circuits.

MAINTENANCE OF THE AQUARIUM	Daily	Weekly	Monthly	Periodically
Check temperature	●			
Check number and health of fish	●			
Check all equipment is in order	●	●		
Partial water change			●	
Check chemical values of water		●	●	
Clean tank bottom		●		
Remove dead leaves		●		
Check air and water tubes				●
Clean tank exterior				●
Clean lamps				●
Clean filter			●	
Restore water level				●

WHEN NECESSARY

Although your little ecosystem under glass should have a long life, it will not last forever. The point will come when the tank will need a general clean-out and rearrangement. By the time this happens, you should be enough of an expert not to be too frightened by the prospect and sufficiently confident to embark on refashioning your aquarium without any difficulty.

TROPICAL AND TEMPERATE FRESH WATERS

Tropical and temperate fresh waters

The watery places of the tropical regions undoubtedly contain the largest number of vegetable and animal forms in the world. To give just a few significant examples, it is estimated that some 2,000 fish species inhabit the basin of the Amazon, with 560 species in the basin of the River Zaire, and no less than 546 in the rivers of Thailand alone. The vast lakes that characterize the Rift Valley of East Africa (Victoria, Tanganyika and Malawi) are home to well over 500 species, many of them endemic. The freshwater habitats of Europe, on the other hand, accommodate only 192 species, and the entire length of the Mississippi a mere 250.

The main reason for this wealth of life is, firstly, favourable climatic conditions, with limited seasonal fluctuations, and secondly, enormous areas of water and hydrographic basins fed by abundant rainfall that in some regions amounts to an average of 1,500 mm (60 in) a year. Nevertheless, of all the parameters to be considered in the context of these watery places, the principal factor has to be temperature. Warmth accounts for the rapidity of the biological and growth processes of very many tropical organisms. Aquatic plants grow all the year round, offering food and shelter to the

fish and invertebrates which, in turn, are linked to one another by complex nutritional relationships. But the rapid vegetational life cycle often leads to the accumulation of organic material which, as it decomposes, diminishes the quantity of dissolved oxygen in the water, this already being reduced by virtue of the high temperatures that militate against the presence of this gas, so indispensable to life. The deoxygenation that is a feature of many tropical habitats with stagnant or feebly flowing water is therefore correlated with the evolution of special respiratory adaptations in many families of tropical fish (Cobitidae, Anabantidae, Clariidae, Callichthyidae, Osteoglossidae, Lepidosirenidae, etc.). Meteorological cycles are in all probability responsible, too, for the extraordinary annual biology of the Cyprinodontidae, whose young seem to appear from nowhere after the rainy season.

The principal tropical areas of interest to aquarists are those biogeographically known as Oriental, Neotropical and Ethiopian, corresponding respectively to Asia, South America and Africa. The two latter continents are of particular zoogeographical importance inasmuch as they were originally parts of Gondwanaland, remaining linked until some 130 million years ago when the supercontinent began to split up. Thus they exhibit both common and exclusive plant and animal communities. Biologically, South America is the kingdom of great tropical rainforests which conceal a dense network of waterways, permanent and temporary, and pools in perfect equilibrium with the surrounding land habitats. This watery world is characterized by a fish population

The black waters of the Amazon basin (below) and the white waters with which they mix (left) are the natural habitats of many of the fish kept in the aquarium.

with astonishing and specific adaptations that include, at one extreme, the capacity to fly, and, at the other, the ability to live by clinging to rocks in the midst of raging cataracts. All of these fish ultimately depend for survival on the organic substances (leaves, fruit, insects, etc.) found in the forest. The wealth of life in these environments is in striking contrast to its poverty when measured on the basis of chemical parameters commonly associated with water productivity. This apparent contradiction is explained by the cycle of matter, here particularly rapid, as is also found in the surrounding forest. Hans Bluntschli, a Swiss scientist who at the beginning of the present century was one of the first to study and understand how the Amazon basin constituted a "harmonious organism," declared that in the River Amazon "there is nothing dead and nothing alive."

Another feature of South America is that its waters are divided into white and black, which is at the origin of certain river names such as Rio Negro. The terms are somewhat misleading. White rivers are, in general, those that rise in the Andes, carrying large quantities of sediment which turns the water cloudy. Black rivers, despite their name, are much clearer, varying in colour from yellow–green to reddish or dark brown. This coloration is due to the presence of humic substances in the soil, which pour into the water especially after rain. These are habitats poor in organic substances, with soft, acidic water; and these characteristics have to be borne in mind when setting up an aquarium and choosing its occupants, in order to reproduce suitable aquatic biotopes.

Africa, on the other hand, despite its huge river basins, such as the Zaire, which covers 3.5 million km^2 (1.3 million sq. miles), is, from the viewpoint of aquarists, the realm of the Cichlidae, found in the great tectonic lakes of the Rift Valley that formed following movements of the earth's crust 10 – 15 million years ago. In these narrow but deep lakes of warm, salt-rich waters, very specialized communities of fish have evolved. More than 200 cichlid species, almost all of them endemic are to be found in Lake Malawi. Also, over 300 species of haplochromine cichlids occur in Lake Victoria, having become differentiated in time as a result of water conditions that

The Asiatic silurid Kriptopterus bicirrhis *only fully reveals its delicate, transparent beauty in the aquarium, and is thus a fascinating and highly valued species.*

humans and fish. This, in fact, is where aquaculture and fish-collecting were both born, the two activities being closely associated. Indeed, these are the regions from which come many of the present-day aquarium fish, even those that are not really Asiatic.

For those of us in the West, however, the most familiar aquatic environments are temperate waters. Fish from these areas, even if wrongfully neglected by some enthusiasts, can provide plenty of interesting and rewarding species. Although far less complex, the rivers and streams of Europe and North America are rewarding habitats worthy of close investigation. Where the terrain permits, it is possible to follow the flow of a river virtually from its source to its mouth. It is particularly rewarding to trace the course of a river that rises in the mountains, with swift, turbulent stretches alternating with those flowing more lazily down gentler slopes, there proceeding even more slowly across the lowland plains, often forming meanders, stagnant branches and pools. Here the water is warm and turbid, rich with vegetation, with beds of fine, soft sediment. This transitional zone courses on until it reaches the sea and hosts species with such specific adaptations, it could provide enough ideas for an aquarium based alone on this habitat.

In fact, in association with this sequence of river habitats is an equally varied succession of fish populations and communities, easily distinguished by virtue of their anatomy, physiology and adaptability. Take, for example, the trout of well-oxygenated cold-water mountain rivers, with its slim, tapering, streamlined body; contrast this with the catfish, carp and sunfish, inhabitants of lakes and slow-coursing rivers where both the temperature and the oxygen levels fluctuate from season to season.

By reason of this correspondence of fish and environment, these rivers may be broadly divided into four zones (frequented respectively by trout, grayling, barber and bream). This breakdown was devised in 1949 by the Belgian hydrobiologist M. Huet, based on the slope and size of the river itself. Obviously, there are no neat separations of the individual zones, particularly in countries where the nature of the terrain is continually changing, as in the Alps, but these subdivisions are generally accepted as valid to this day.

have remained virtually unchanged in 500,000 years. Such cases of specialization can provide excellent possibilities for stocking aquaria with single species.

The watery habitats of the tropical regions of Asia are much more varied. They are home to imported fish such as the *Macropodus* species and the goldfish, capable of living in unheated aquaria. Here we find slow, turbid lowland rivers as well as rivers hidden away in impenetrable forests, rushing mountain streams, temporary pools, canals and rice paddies, these last being man-made yet inhabited by a variety of fish species. The juxtaposition of natural and artificial waterways is not surprising, for in Asia there is a particularly strong link between

ACANTHOPHTHALMUS SEMICINCTUS Fam. Cobitidae

Common name
Half-banded coolie loach.
Dimensions
Up to 8 cm (3 in).
Distribution
Southeast Asia.
Food
In aquarium: prefers benthic organisms such as *Tubifex* and insect larvae. Eats dried fish food and detritus.
General care
Adapts to a wide range of water conditions but must have suitable hiding places because it spends much of its time concealed. Tolerates various temperatures but 21° – 25°C (70° – 77°F) is recommended.
Notes
Cure of skin lesions and fungal infections is difficult. These loaches seldom breed in captivity and there are only conflicting reports on success rates.

Acanthophthalmus semicinctus

AEQUIDENS CURVICEPS Fam. Cichlidae

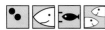

Common name
Flag cichlid.
Dimensions
Up to 8 cm (3 in).
Distribution
Amazon basin.
Food
In nature: generally invertebrates. In aquarium: live food, dried fish food.
General care
Tranquil species that lives happily with others. Several specimens can be kept in a 55 – liter (12 – gal) tank. Tolerates plants. Cover the rear wall and part of the others and create hiding places. Maintain temperatures of 18° – 27°C (65° – 80°F).
Notes
The male has longer, more pointed dorsal and anal fins. Eggs are laid on the substrate. Both parents tend eggs and larvae. Another popular species is *A. pulcher*.

Aequidens curviceps

ALESTES LONGIPINNIS Fam. Characidae

Common name
Long-finned characin.
Dimensions
Up to 15 cm (6 in).
Distribution
Tropical West Africa.
Food
In nature: principally flying insects. In aquarium: fruit flies, meat flies, dried fish food, other insects.
General care
A shy species which needs water that is not too hard and slightly alkaline in which to breed. Newly hatched larvae readily accept *Artemia* nauplii. It is best to remove adults from the aquarium to avoid the eggs and fry being eaten. Use a large 400–liter (88–gal) aquarium and keep several fish together. Create hiding places with plants, stones and roots but leave free space, too.
Notes
In nature these fish prefer clear waters and are mainly found in Sierra Leone and Zaire. Mature males are distinguished by long dorsal fins, usually red-tinted. Females have standard-length yellowish dorsal fins. They lay eggs in open water close to plants. The fertilized eggs, approximately 2 mm in diameter, drop to the bottom and hatch after 6 days. This species is sometimes placed in the family Alestidae.

ANABAS TESTUDINEUS Fam. Anabantidae

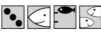

Common name
Climbing perch.
Dimensions
Up to 25 cm (10 in).
Distribution
India to Sri Lanka and Southeast Asia, China and Philippines.
Food
In nature: carnivorous. In aquarium: fish foods such as raw meat, tablets and earthworms.
General care
Use a well-covered and only partially filled tank, with vegetation and a branch or stone above water level. The fish have to breathe air and suffocate if they remain submerged. No restrictions on water quality. Keep temperatures at 20° – 30°C (68° – 86°F). It is best to use an undergravel filter.
Notes
This species has no particular markings, and grows quite big for an aquarium fish. It is of great interest because of its ability to climb out of the water and crawl about. *A. testudineus* can, in fact, survive for several hours out of water. It is timid but may turn quarrelsome with its aquarium companions.

Alestes longipinnis

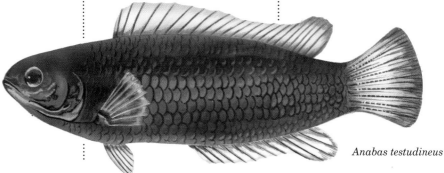

Anabas testudineus

38 • TROPICAL FRESH WATERS

ANABLEPS ANABLEPS Fam. Anablepidae	**ANOPTICHTHYS JORDANI** Fam. Characidae	**ANOSTOMUS ANOSTOMUS** Fam. Anostomidae	**APHYOSEMION AUSTRALE** Fam. Cyprinodontidae
			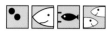

ANABLEPS ANABLEPS
Fam. Anablepidae

Common name
Four-eyed fish.
Dimensions
Up to 30 cm (12 in).
Distribution
Muddy zones of northeastern South America.
Food
In aquarium: ideally insects, but readily accepts artificial foods provided they float for a while on the surface.
General care
Provide a large, shallow and well-covered tank, with sand and mud. Fill with brackish water at 24° – 28°C (75° – 82°F). This species will breed in captivity. Eggs develop inside the mother's body and twice yearly a few young (on average 6) are born; they are already 3 – 4 cm (1¼ – 1½ in) long and able to feed immediately on *Artemia* nauplii. Adults may show aggression towards other species.
Notes
In nature these fish like to rest on mud or sand or sand beds lightly washed by water. For this reason it is advisable for the substrate to protrude above the surface in some places.

ANOPTICHTHYS JORDANI
Fam. Characidae

Common name
Blind cavefish.
Dimensions
Up to 9 cm (3½ in).
Distribution
Caves of central-eastern Mexico.
Food
In nature: various kinds of small animals. In aquarium: flake and freeze-dried fish food.
General care
A shy species which will live better in groups of its own kind, in a spacious tank. It needs no particular water quality and adapts to temperature fluctuations, ideally between 18° and 24°C (65° and 75°F). To create an environment similar to that of the natural habitat, place cave-like structures in the aquarium.
Notes
Young cavefish have poorly defined eyes which disappear with age, the sockets becoming filled with tissue. Specimens living in caves are unpigmented and thus look pink. If kept in a lighted place they become silvery. They are relatively easy to look after. Fertilized eggs hatch within 3 – 4 days of being laid, and the larvae start to swim and feed after about 6 days. Large-mouthed larvae immediately accept *Artemia* nauplii. This species is now known also as *Astyanax mexicanus*.

ANOSTOMUS ANOSTOMUS
Fam. Anostomidae

Common name
Striped anostomus
Dimensions
Up to 14 cm (5 ½ in).
Distribution
Regions of Guyana and Amazon basin.
Food
In nature: omnivorous. In aquarium: adapts to any kind of fish food containing vegetable matter; also worms.
General care
A large aquarium of at least 75 liters (16½ gal) is needed to allow maximum growth, preferably with many plants and diffused lighting. No restrictions on water quality. Recommended temperatures are 24° – 28°C (75° – 82°F).
Notes
Like many other species of the family Anostomidae, this anostomus has the strange habit of swimming with its body in a sideways position with head turned downward. Males are smaller and more colourful than females. The related species *A. trimaculatus* is more robust and displays three black spots on the flanks.

APHYOSEMION AUSTRALE
Fam. Cyprinodontidae

Common name
Lyretail.
Dimensions
Up to 5 cm (2 in).
Distribution
Equatorial West Africa.
Food
In aquarium: dried fish food mixed once a week with live food such as fruit flies, ants, mosquito larvae, *Daphnia*, etc.
General care
Lyretails are normally kept in small aquaria of 7 – 12 liters (1½ – 2½ gal), with only one male and one or more females in each tank. Aeration and heating are not usually necessary. Temperatures of 20° – 25°C (68° – 77°F) are sufficient. The water should be softish and slightly acid, the substrate of boiled peat. Plants are unnecessary but may be useful as hiding places for the females who would otherwise be courted by the male to the point of exhaustion.
Notes
This is one of the numerous *Aphyosemion* species successfully bred by aquarists. The females lay 10 – 12 eggs each day which adhere to the vegetation close to the surface. They hatch within about two weeks and the fry feed on *Artemia* nauplii. The male is more brightly coloured than the female.

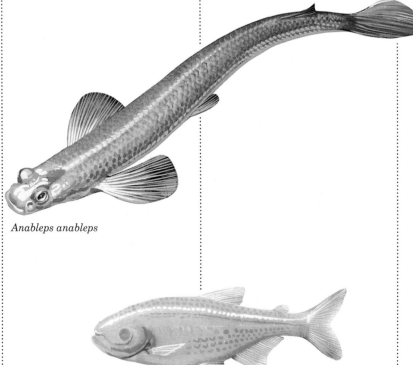

Anableps anableps

Anoptichthys jordani

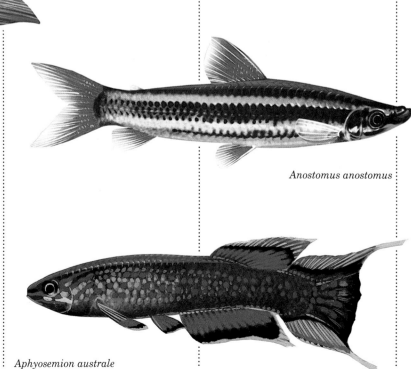

Anostomus anostomus

Aphyosemion australe

TROPICAL FRESH WATERS • 39

APHYOSEMION SJOESTEDI
Fam. Cyprinodontidae

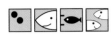

Dimensions
Up to 12 cm (5 in).
Distribution
Eastern Nigeria and Cameroon.
Food
In nature: probably insects. In aquarium: normally requires live food.
General care
A quarrelsome species. Keep a single male and several females in a 40-liter (8¾-gal) planted aquarium, with or without substrate, at 17° – 24°C (63° – 75°F). Neither aeration nor filtration are needed.
Notes
The female lays eggs on the substrate; if there is none on the floor of the tank, install suitable supports. The fish should be encouraged to eat live food to increase the number of eggs, which may be as many as 20 a day. The eggs can be incubated in the water (for around 30 days) or in moist peat moss for perhaps twice as long. The newly hatched larvae feed on *Artemia* nauplii.

APISTOGRAMMA AGASSIZI
Fam. Cichlidae

Common name
Agassiz' dwarf cichild.
Dimensions
Up to 8 cm (3 in).
Distribution
Amazon basin to River Paranà and Paraguay.
Food
In nature: mainly carnivorous. In aquarium: live food such as small crustaceans, worms and insect larvae.
General care
The water should be slightly acid, not hard, and filtered through peat. Keep several females with a single male in a large tank. Provide plenty of hiding places, such as rock caves or plants. Recommended temperatures are 21° – 25°C (70° – 77°F).
Notes
The male of the species lives territorially with several females. Spawning takes place inside a cave, in a spot prepared by the male. The female cares for the eggs and larvae while the male patrols the territory.

ASTRONOTUS OCELLATUS
Fam. Cichlidae

Common name
Oscar, velvet cichlid.
Dimensions
Up to 30 cm (12 in).
Distribution
Northern areas of South America.
Food
In nature: predator of small fish and insects. In aquarium: supply a varied diet comprising crustaceans, worms, meat and insects. Normally they adapt to stick and pellet fish foods.
General care
Keep in a single-species tank. Small specimens can live in a 72 liter (15¾-gal) aquarium but since adults may reach 30 cm (12 in), a 200 liter (44 gal) tank is preferable. Although sometimes shy in captivity, most individuals are relatively tranquil. They live comfortably at temperatures of 20° – 27°C (68° – 80°F). No restrictions on water quality, but thorough filtration is necessary. Because the fish are tireless burrowers, plants are not recommended. Use wood and stones as hiding places and fill the bed with gravel. They can be bred in captivity.
Notes
Coloration varies a good deal according to forms available: these also include the red oscar. Females lay eggs in the open, normally on a smooth rock. These hatch in 3 – 4 days and the larvae are placed in a shallow trench until they are free-swimming (5 – 6 days); at this point the parents are better separated. Feed the larvae with *Artemia* nauplii.

ASTYANAX MEXICANUS
Fam. Characidae

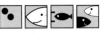

Dimensions
Up to at least 9 cm (3½ in).
Distribution
From Texas to Panama.
Food
In nature: small organisms. In aquarium: they accept a broad variety of natural or dried fish foods.
General care
Shy species. Keep in a planted aquarium. Use roots as hiding places and to make the aquascape more authentic. Room temperatures of 16° – 22°C (61° – 72°F) are satisfactory, but temperatures closer to those of the original habitats are better still.
Notes
There is a blind form of this species living in the Cueva Chica caves of San Luis Potosí (Mexico). It was once thought to be a distinct species, but more detailed research and the ease with which the blind and normal forms interbreed have shown beyond doubt that they are both the same species.

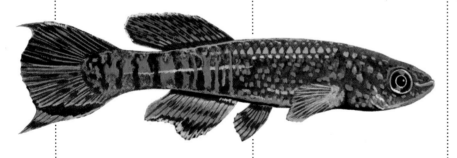

Aphyosemion sjoestedi

Apistogramma agassizi

Astronotus ocellatus

BARBUS CONCHONIUS
Fam. Cyprinidae

Common name
Rosy barb.
Dimensions
Up to 14 cm (5½ in).
Distribution
India.
Food
In nature: omnivorous. In aquarium: dried fish food mixed with live food.
General care
It adapts to life both in large and small aquaria at a temperature of about 22°C (72°F). No restrictions on water quality, but the bed must be made up of soft, peaty materials because the fish likes to burrow in the sediment in search of food. It is easy to breed but after the eggs are laid, among the plants, they must be protected from the parents who tend to eat them.
Notes
This is a very popular fish with aquarists and particularly beautiful in a well-planted aquarium with the bed made up of dark red material.

BARBUS CUMINGI
Fam. Cyprinidae

Common name
Cuming's barb.
Dimensions
Up to 5 cm (2 in).
Distribution
Sri Lanka.
Food
In nature: omnivorous. In aquarium: fish food containing vegetable matter; mix with live food, *Daphnia*, worms, insects, etc.
General care
A hardy, tranquil species which can live in a small 40-liter (8¾-gal) aquarium. The water should preferably be softish and slightly acid, at a temperature of 24° – 27°C (75° – 80°F). Keep a number of them together.
Notes
From the mountain forest streams of Sri Lanka, these fish are strong swimmers. They need sufficient space for moving around and tend to settle in the lower half of the tank. The eggs, normally laid among the vegetation, hatch within about 25 hours and the fry swim freely after a few days.

BARBUS NIGROFASCIATUS
Fam. Cyprinidae

Common name
Black ruby barb.
Dimensions
Up to 5 cm (2 in).
Distribution
Sri Lanka.
Food
As for *Barbus cumingi*.
General care
As for *B. cumingi*, but it adapts to slightly higher temperatures. Mature males become more colourful than females.
Notes
It is one of the sturdier *Barbus* species and is particularly suitable for beginners. As in the previously described species, the bed should consist of soft materials. The vegetation need not be too dense and it is advisable to include floating plants which attenuate the strong light that these fish dislike.

BARBUS PENTAZONA
Fam. Cyprinidae

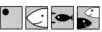

Common name
Banded barb.
Dimensions
Up to 5 cm (2 in).
Distribution
Southeast Asia.
Food
In nature: omnivorous. In aquarium: dry foods containing vegetable matter; mix with live food.
General care
Hardy and fairly tranquil species. Keep several of them together in aquaria of 40 liters (8¾ gal) or more, in fairly soft water, at temperatures of 20° – 27°C (68° – 80°F).
Notes
The classification of this species, actually a subspecies, has often been modified, so that from time to time various authors have ascribed it to the genera *Barbodes*, *Capoeta* and *Barbus*, and its common name has been given to several species simultaneously, as for example *B. hexazona* and *B. oligolepis*. For present purposes the simplified nomenclature has been used, although the more accurate name is *Barbus tetrazona partipentazona*.

Astyanax mexicanus

Barbus cumingi

Barbus conchonius

Barbus nigrofasciatus

Barbus pentazona

BARBUS SCHWANENFELDI
Fam. Cyprinidae

Common name
Schwanenfeld's barb.
Dimensions
Up to 35 cm (14 in).
Distribution
Sumatra, Borneo, Thailand.
Food
In nature: omnivorous; its diet comprises a good deal of plant matter. In aquarium: dried fish food mixed with lettuce and other vegetables.
General care
It is an excellent aquarium species but needs a large tank. While adapting well to life with other species, it should not be put in with smaller specimens. A dark background of peat is recommended. Use only sturdy plants. No restrictions on water quality. Maintain temperatures of 21° – 26°C (70° – 79°F).

BARBUS TETRAZONA
Fam. Cyprinidae

Common name
Tiger barb.
Dimensions
Up to 7 cm (2¾ in).
Distribution
Sumatra and Borneo.
Food
In nature: omnivorous. In aquarium: it adapts to fish food containing vegetable matter; mix with live food.
General care
Active species which tends to bite its aquarium companions. Keep a number of them together in a relatively large tank of 60 liters (13 gal) or more, with fairly soft water and at temperatures of 20° – 27°C (68° – 80°F).
Notes
In this case simplified nomenclature has been used, this fish usually being classified by the more precise name of *Barbus tetrazona tetrazona*.

Barbus tetrazona

BETTA SPLENDENS
Fam. Belontiidae

Common name
Siamese fighting fish.
Dimensions
Up to 6 cm (2½ in).
Distribution
Thailand, Indochina, Malaysia.
Food
In nature: a wide variety of small organisms. In aquarium: normal fish food, including live food.
General care
A single male can be kept with one or two females in an aquarium of 40 liters (8¾ gal) or more, alone or with other species. There are no restrictions on water quality, but breeding seems to be stimulated by a low water level of 15 cm (6 in) or less. Introduce floating plants and keep at temperatures of 26° – 30°C (79° – 86°F).
Notes
Betta splendens builds bubble nests and will breed in captivity. After courtship, which involves mutual rubbing, blows and circling movements, the male wraps himself around the female. The eggs hatch within 24 – 35 hours; the larvae accept rotifers and infusorians after beginning to swim freely, at which point they must be separated even from the adults. Artificial selection has led to the obtaining of a broad range of magnificently coloured varieties with ever more elaborate fins.

BOTIA MACRACANTHA
Fam. Cobitidae

Common name
Clown loach.
Dimensions
Up to 30 cm (12 in) in nature, but rarely exceeding 15 cm (6 in) in captivity.
Distribution
Sumatra and Borneo.
Food
In nature: detritus organisms. In aquarium: dried fish food and worms, especially *Tubifex*.
General care
This fish has predominantly nocturnal habits, preferring to rest during the daytime in the shelter of vegetation or beneath stones and roots. If threatened, it emits snapping sounds. It is best to keep three or more individuals together. The water should not be too hard, and the aquarium should contain plants and have a soft substrate. Recommended temperatures are 22° – 26°C (72° – 79°F). Cover the tank.
Notes
The clown loach is long-lived in the aquarium but grows slowly, never reaching the dimensions found in nature. This may explain why there is so little information about its breeding habits in captivity, for some authors believe, in fact, that it grows too slowly to attain sexual maturity in aquarium conditions.

Botia macracantha

Barbus schwanenfeldi

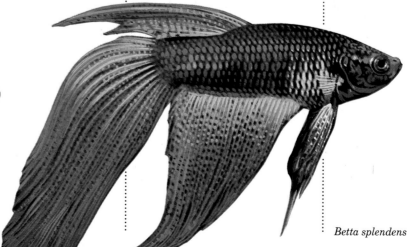

Betta splendens

BRACHYDANIO ALBOLINEATUS
Fam. Cyprinidae

Common name
Pearl danio.
Dimensions
Up to 5 cm (2 in).
Distribution
Southeast Asia.
Food
Accepts both dried and freeze-dried fish food, but it is best to mix this diet with worms and fresh shrimp.
General care
The aquarium should have plenty of plants and a cover to prevent these fish, which swim in the middle and upper layers of the aquarium, leaping out of the tank. The water should be soft or slightly hard, even during the reproductive season. Morning light seems to stimulate breeding, which lasts several hours. Females lay up to 600 eggs which must be protected from the parents. They hatch within about 24 hours and the fry swim freely in the tank after 6 – 7 days.
Notes
This species has been known to aquarists since 1911. Seen in reflected light, the fish assumes a violet or iridescent blue coloration, whereas in transmitted light it appears green.

BRACHYDANIO RERIO
Fam. Cyprinidae

Common name
Zebra danio.
Dimensions
Up to 5 cm (2 in).
Distribution
India.
Food
In aquarium: it adapts to any form of aquarium diet.
General care
Extremely easy to keep. An aquarium with plenty of plants and free space for swimming is recommended; beginners can start with plastic plants. The fish tolerates temperatures of 16° – 38°C (61° – 100°F). Males are smaller and slimmer than females and become more colourful. For breeding, choose a mature female and place her in a suitable tank of 20 – 40 liters (4¼ – 8¾ gal) maintained at 22° – 24°C (72° – 75°F), with strong, small-leaved plants. Take two active, brightly coloured males and put them in with the female. Make sure the fish are fed frequently with a nutritious diet, preferably mixing in live food. The females lay on the vegetation and should be removed at once so that they do not eat the eggs. The larvae hatch in 1 – 2 days. Feed them with rotifers, infusorians and other small items, including dried fry fish food. After a few days, they should take *Artemia* nauplii.
Notes
This is an ideal fish for novices but should not be neglected by experts, particularly since it adapts well to living with others.

BRACHYGOBIUS XANTHOZONA
Fam. Gobiidae

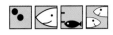

Common name
Bumblebee fish, wasp goby.
Dimensions
Up to 4 cm (1½ in).
Distribution
Indonesia.
Food
In nature: benthic invertebrates. In aquarium: it accepts a wide variety of live foods but also adapts to good dried fish food.
General care
A timid fish, best kept in a small tank reserved for this species. Use brackish water, either diluted sea water (20% sea water) or fresh water with added sea salt (2 tsp salt to every 10 liters [2 gal]). Cover the tank bottom with sand, provide stones for shelter and, if desired, plants that tolerate salinity. The species is difficult to breed. The eggs are attached to the ceiling and walls of the shelter and are watched over by the male. The female should be removed. The larvae require infusorians and other small live food items. Recommended temperatures are 24° – 28°C (75° – 82°F)
Notes
The typical yellow coloration with four broad black bands tends to be modified with age. In fact, in older specimens the dark bands are reduced in size. In addition to tropical forms, the family Gobiidae contains many temperate freshwater species with very similar habits, although their colours are less conspicuous.

CALLICHTHYS CALLICHTHYS
Fam. Callichthyidae

Common name
Mailed catfish.
Dimensions
Up to 18 cm (7 in).
Distribution
Tropical South America.
Food
In nature: omnivorous, feeding avidly on benthic organisms. In aquarium: insects, worms and fish food that falls to the bottom.
General care
Hardy, tranquil species. Create hiding places and use sturdy plants. No restrictions on water quality. Keep at temperatures of 21° – 26°C (70° – 79°F).
Notes
This species adapts readily to life with others because it is placid and has cleaning habits. In nature it lives in shallow water. It has an accessory intestinal respiratory apparatus which allows it to breathe air when the quantity of dissolved oxygen in the vicinity is low. It has been bred in captivity. The male builds an air-bubble nest among the plants near the surface. After the female lays her eggs, he guards the nest. The larvae need dried fish food in powdered form.

Brachydanio albolineatus

Brachygobius xanthozona

Brachydanio rerio

Callichthys callichthys

TROPICAL FRESH WATERS • 43

CARNEGIELLA STRIGATA
Fam. Gasteropelecidae

Common name
Marbled hatchetfish.
Dimensions
Up to 4.5 cm (1¾ in).
Distribution
Guyana, Surinam, Amazon basin.
Food
In nature: small insects. In aquarium: it needs insects to thrive but may also adapt to some forms of dried fish food.
General care
Keep at least six specimens in as long an aquarium as possible. Cover with a suitable lid, leaving a space of at least 8 cm (3 in) between this and the surface. This fish prefers softish, peaty water, maintained at a temperature of 23° – 30°C (74° – 86°F).
Notes
It is a placid fish with an active period at dusk. It prefers diffused lighting rather than bright illumination, which makes its colours fade. Breeding in captivity is rare. According to some observations, it seems that the male courts the female by first swimming around her before the pair head for the vegetation where she lays eggs that adhere to the plants and hatch within about 30 hours. The fry, which at first remain near the bottom of the tank, take on the adult form after some 20 days.

CHALCEUS MACROLEPIDOTUS
Fam. Characidae

Common name
Pink-tailed characin.
Dimensions
Up to 25 cm (10 in).
Distribution
Guyana.
Food
In nature: carnivorous. In aquarium: fish and minced meat, earthworms and most types of dried fish food.
General care
Although a hardy species, it needs a large aquarium of 400 liters (88 gal) or more in order to thrive. This must be well covered to prevent the fish leaping out. If possible, keep six or more individuals together in a well-planted tank. Recommended temperatures are 24° – 26°C (75° – 79°F).
Notes
The large mirror-like scales and the tail which veers from red to pink make this a particularly prized aquarium fish. It is a tranquil species with fish of its own size. In nature, the fish form groups and are caught with fixed nets.
It can be bred in captivity by transferring specimens which seem ready to mate into an aquarium with bushy plants, on which the eggs are laid. These, some 3,000 in number, should be placed in a suitable tank where they will hatch within 48 hours at 27°C (80°F). The young begin to swim and feed after about 4 days. Initially they can be given *Artemia* nauplii but as they grow they will need larger items such as fish, worms and dried fish food.

CHANDA WOLFII
Fam. Centropomidae

Dimensions
Up to 12 cm (4 in) in captivity, 20 cm (8 in) in nature.
Distribution
Indonesia, Thailand.
Food
In nature: small planktonic organisms. In aquarium: needs live food such as *Daphnia*, small crustaceans, mosquito larvae, moths, etc.
General care
Suitable for keeping with other tranquil species. Provide with small-leaved plants to serve as hiding places. Recommended temperatures are 22° – 29°C (72° – 84°F).
Notes
The smaller *Chanda ranga* is certainly better known to aquarists and, unlike *C. wolfii*, is fairly easy to breed in captivity.

CHEIRODON AXELRODI
Fam. Characidae

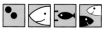

Common name
Cardinal tetra.
Dimensions
Up to 5 cm (2 in).
Distribution
Higher reaches of Rio Negro.
Food
As for *Paracheirodon innesi*.
General care
Keep in groups of six or more individuals, in water that is not too calcareous and slightly acid, at temperatures of 22° – 27°C (72° – 80°F). Fill tanks with small-leaved plants. Difficult to breed.
Notes
This is the most colourful of all the Characidae. When discovered in the 1950s, it created a sensation in the fish-keeping world. It is also known as *Paracheirodon axelrodi* by some authorities.

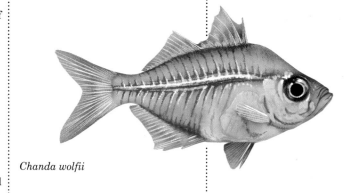

Chanda wolfii

Cheirodon axelrodi

Carnegiella strigata

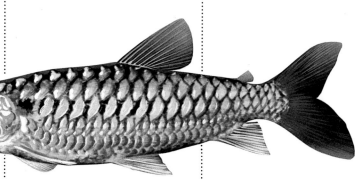

Chalceus macrolepidotus

CICHLASOMA MEEKI
Fam. Cichlidae

Common name
Firemouth cichlid.
Dimensions
Up to 15 cm (6 in).
Distribution
Central and South America.
Food
In aquarium: ordinary fish food.
General care
Use an aquarium of at least 75 liters (16 gal) and, if desired, include hardy plants. Do not as a rule disturb the bed, except when the fish are spawning. Temperatures of 20° – 26°C (68° – 79°F) are recommended. Introduce flat stones for egg laying. After the eggs are laid on the substrate, both parents watch over them and care for the larvae, which accept *Artemia* nauplii as soon as they begin to swim freely. Larvae and adults may be left together or separated, as desired.
Notes
During the spawning season, the males fight one another to defend territory. These combats involve threatening the rival by enlarging the gill covers to their full extent. The species is sensitive to rapid fluctuations of temperature. Sexual dimorphism is normally not very evident. By observing the fish carefully, however, it is possible to identify males by their pointed dorsal and anal fins and the females by their less robust body structure. The size for initial spawning is about 8 cm (3 in). Recent studies put this, and a number of related cichlids, in the genus *Heros* or *Thorichthys*.

CICHLASOMA OCTOFASCIATUM
Fam. Cichlidae

Common name
Jack Dempsey.
Dimensions
Up to 20 cm (8 in).
Distribution
Amazon and Rio Negro basins.
Food
In nature: omnivorous. In aquarium: accepts and adapts to conventional diet.
General care
Usually aggressive. It is best to keep it alone or in pairs. Tends to uproot plants for spawning. No restrictions on water quality and temperatures around 24°C (75°F). The species is rather sensitive to old water, so it is advisable to change a quarter of the water in the tank every 3 – 4 weeks. Give plenty of aeration but not too much as, given the burrowing habits of the species, it is easy to cloud the water.
Notes
This celebrated aquarium fish becomes particularly aggressive in the spawning season. When sexually mature, the male takes on deeper colours and develops fatty deposits on the head. His dorsal and anal fins are more pointed. Both parents busy themselves cleaning the stones or pieces of wood on which the eggs are laid, and then care for both the eggs and the larvae. The species is easy to breed.

CLARIAS BATRACHUS
Fam. Clariidae

Common name
African catfish.
Dimensions
Up to 55 cm (22 in).
Distribution
Southeast Asia, introduced to Florida.
Food
In nature: omnivorous. In aquarium: any type of fish food.
General care
Shy but voracious species. Only young specimens are suitable for home aquaria. Maintain a temperature of 22° – 27°C (72° – 80°F) in a tank with plenty of hiding places. Plants should be put in pots.
Notes
This is not a species to be recommended for a home aquarium although it is hardy and today widely distributed. It has been introduced to Florida, where it can sometimes be seen slithering through fields, because it has a supplementary respiratory system that enables it to stay out of water for long periods.

COLOSSOMA BRACHYPOMUM
Fam. Characidae

Common name
Pacu.
Dimensions
Up to 60 cm (24 in).
Distribution
Amazon basin, Guyana.
Food
In nature: vegetarian, but also eats seeds, fruit and plant matter. In aquarium: fish food in pellet form mixed with wheat grains and other plant material. It also takes meat.
General care
This species grows very big and therefore needs an aquarium of at least 750 liters (165 gal) at temperatures of 20° – 25°C (68° – 77°F).
Notes
Pacu are edible fish widely eaten by the people of Amazonas and Guyana. Little is known of their breeding behaviour but there are indications that they lay eggs in holes dug in the substrate. Some species migrate from the larger rivers into smaller tributaries in the spawning season.

Clarias batrachus

Cichlasoma meeki

Colossoma brachypomum

Cichlasoma octofasciatum

TROPICAL FRESH WATERS • 45

COPEINA ARNOLDI
Fam. Lebiasinidae

Common name
Splashing tetra.
Dimensions
Up to 8 cm (3 in).
Distribution
Lower Amazon basin.
Food
In nature: principally insects which live in or fall on to the water. In aquarium: insects, dried fish food.
General care
They prefer an aquarium of 75 liters (16 gal) or more with plenty of vegetation and emergent plants that provide an area for egg laying. Maintain temperatures of 24° – 26°C (75° – 79°F).
Notes
Spawning occurs on the underside of plant leaves above the water surface. Having selected a suitable place, both male and female leap from the water and collide with the leaf, sticking to it with their fins, which act as suckers. In these brief moments the female lays a number of eggs which are immediately fertilized by the male. The operation is repeated until a hundred or so eggs are laid. At this point the male remains alone beneath the cluster of eggs, using his caudal fin to splash them regularly with water. The fry hatch within 36 – 48 hours and fall straight into the water. This species is also known as *Copella arnoldi*.

CORYDORAS AENEUS
Fam. Callichthyidae

Common name
Bronze catfish.
Dimensions
Up to 7 cm (2¾ in).
Distribution
Trinidad, South America to basin of River Plate.
Food
In nature: omnivorous. In aquarium: worms, insects and dried fish food containing vegetable matter.
General care
A robust, not particularly elegant but tranquil species, whose only problem is that it uproots plants. It lives happily with other species. Maintain temperatures of 19° – 27°C (67° – 80°F). A well-planted aquarium with a trench is necessary for breeding. The eggs are laid on plant leaves and hatch within about 6 days. The larvae swim after 8 days and may be fed with *Artemia* nauplii.
Notes
The species acclimates well to tanks containing a number of fish. It is quite active and spends much of the time circling the aquarium, touching the bottom and the walls with its barbels. Like other *Corydoras* species, it is a good aquarium fish.

CYPRINODON MACULARIS
Fam. Cyprinodontidae

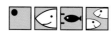

Dimensions
Up to 7 cm (2¾ in).
Distribution
California, Arizona, Lower California, Mexico.
Food
In nature: omnivorous, principally small crustaceans. In aquarium: any kind of dried fish food.
General care
A very robust species which lives at temperatures of 9° – 45°C (48° – 113°F) in water of about 7% salinity.
Notes
It lives in conditions that are unsuitable for other fish.

DATNIOIDES MICROLEPIS
Fam. Lobotidae

Dimensions
Up to 40 cm (16 in).
Distribution
Southeast Asia from India to New Guinea.
Food
In nature: carnivorous. In aquarium: start, if necessary, with live fish, then go on to whole dead fish, and finally minced mussels.
General care
It will adapt to living in a tank of 400 liters (88 gal) or more with other large tranquil species. Use stones and submerged wood to create hiding places. Plants are useful but leave sufficient free space. There are no restrictions on water quality, but many aquarists prefer to use slightly brackish water, for example 20 g of salt to very 10 liters (2 gal) of water (2 parts per thousand).

Cyprinodon macularis

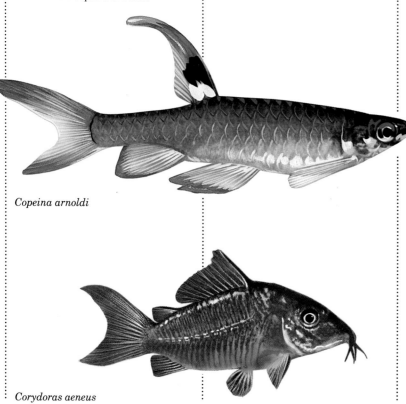

Copeina arnoldi

Corydoras aeneus

Datnioides microlepis

ETROPLUS MACULATUS
Fam. Cichlidae

Common name
Orange chromide.
Dimensions
Up to 10 cm (4 in).
Distribution
India and Sri Lanka.
Food
In nature: omnivorous. In aquarium: include live food and vegetable matter in diet.
General care
Lives in brackish water, so add sea salt or sea water to fresh water to maintain a specific gravity of 1.005 – 1.025. Use an aquarium with plants and a sandy bed, with flat stones set vertically and submerged wood. Keep temperatures at 21° – 28°C (70° – 82°F).
Notes
The eggs, black in colour, adhere to the stones on the bottom by means of a thin stalk. Both parents watch over the eggs and the fry which are born in 5 – 6 days. The newborn fish affix themselves to the flanks of the parents for a while, then leave to swim freely.

EXODON PARADOXUS
Fam. Characidae

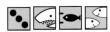

Dimensions
Up to 15 cm (6 in).
Distribution
Northeastern areas of South America; Rio Rupununi (Guyana), Rio Branco (Brazil).
Food
In nature: stomach contents indicate that *Exodon* feeds mainly on the scales of other fish. In aquarium: live food, including small fish, is readily accepted.
General care
It is an active jumper, and by reason of its aggressive nature it must be kept in a tank with individuals of the same species or of robust related species. The large tank should be accessible to diffused sunlight. The temperature must be maintained at 24° – 28°C (75° – 82°F).
Notes
This is one of the loveliest and most active characins. Spawning is vigorous; the large eggs are laid on plants and hatch within a day or so. The species is problematical to breed and rear, and because of its aggressive habits, it is difficult to keep in an aquarium. Nevertheless, for the expert aquarist this provides an interesting challenge.

FUNDULUS CHRYSOTUS
Fam. Cyprinodontidae

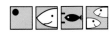

Common name
Golden ear.
Dimensions
Up to 8 cm (3 in).
Distribution
South Carolina to Florida.
Food
In aquarium: adapts rapidly to dried and frozen fish foods.
General care
Easy to breed. Maintain temperatures of 22° – 26°C (72° – 79°F). At time of spawning, introduce two or three females for each male. The eggs are laid among vegetation made up principally of soft-leaved (*Myriophyllum*-type) plants. It is advisable to transfer the eggs to an incubation tank or to remove the parents, who will readily feed on them.
Notes
If kept together with other livelier fish, this species tends to remain in a subordinate position and not to develop its typical coloration, namely iridescent olive – green for males, grey – brown for females. They are best kept in a tank on their own.

GEOPHAGUS JURUPARI
Fam. Cichlidae

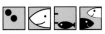

Common name
Earth-eater.
Dimensions
Up to 24 cm (9½ in).
Distribution
Northern areas of South America.
Food
In nature: rakes the bottom for invertebrates or other organic matter. In aquarium: fish foods containing vegetable matter, mixed, if necessary, with live foods.
General care
Use an aquarium of at least 75 liters (16½ gal) with a sandy bed and flat stones. Hardy plants can also be introduced. No restrictions on water quality. Maintain temperatures of 21° – 27°C (70° – 80°F). An outside filter is advisable because these fish are continually burrowing.
Notes
The species has interesting breeding habits. Eggs are laid on and adhere to a flat rock. After a couple of days the female takes the eggs into her mouth and incubates them. Remove the male when the eggs hatch. The female continues to rear the young for some time. These accept *Artemia* nauplii as soon as they become free-swimming.

Etroplus maculatus

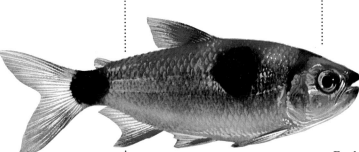

Exodon paradoxus

Fundulus chrysotus

Geophagus jurupari

TROPICAL FRESH WATERS • 47

GNATHONEMUS PETERSI
Fam. Mormyridae

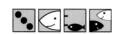

Common name
Elephant fish.
Dimensions
Up to 23 cm (9 in).
Distribution
Tropical Africa, including Cameroon and Zaire.
Food
In nature: small organisms living in mud. In aquarium: small worms, insect larvae and plankton (*Daphnia, Cyclops*).
General care
There are no restrictions on water quality, but the temperature must be 20° – 28°C (68° – 82°F), the aquarium dimly lit, with surface plants, rocks and bits of wood to make hiding places. Food must be abundant. It tolerates other fish and can be kept in the aquarium with associated species. However, it may react against other electric fish.
Notes
This strange fish is probably the best known of the Mormyridae. It adapts well to captivity but care must be taken to provide it with enough live food.

GYMNOCORYMBUS TERNETZI
Fam. Characidae

Common name
Black tetra, black widow.
Dimensions
Up to 5 cm (2 in).
Distribution
Rio Paraguay, Rio Negro.
Food
In nature: insects and other small organisms. In aquarium: dried fish food, insects.
General care
Can be kept in small aquaria. No restrictions on water quality. Maintain temperatures of 24° – 28°C (75° – 82°F). Decorate with rocks and plants. Easy to breed.
Notes
Gymnocorymbus ternetzi is bred commercially by specialists and offered for sale. It is a recommended species for beginners. Observed by back lighting, the female's abdominal cavity appears rounded, the male's pointed. The males, moreover, are recognizable by the light spots on the caudal fin. This species is reared in small groups which normally tend to remain in the middle and upper layers of the aquarium.

GYMNOTUS CARAPO
Fam. Gymnotidae

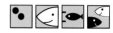

Dimensions
Up to 60 cm (24 in).
Distribution
From Guatemala to Argentina.
Food
In nature: carnivorous. In aquarium: fish, lean ox heart and worms.
General care
Easy to keep but aggressive towards others of its kind. A dark aquarium of 400 liters (88 gal) or more should be decorated with hollow trunks and roots. Maintain temperatures of 23° – 28°C (74° – 82°F). No restrictions on water quality.
Notes
Although the weak electric shocks emitted by these fish will not disturb the other aquarium occupants too much, the species is best kept in a separate tank of a size proportionate to the numbers. This is a prevalently nocturnal species which may be induced to appear during the day provided the aquarium is kept dark.

HAPLOCHROMIS BURTONI
Fam. Cichlidae

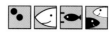

Dimensions
Up to 12 cm (5 in).
Distribution
Lake Tanganyika and associated rivers.
Food
In nature: carnivorous. In aquarium: prefers live food, worms, etc., but can manage with dried fish food.
General care
It can be kept in an aquarium with a number of species, but in the spawning season the pairs should be isolated in a tank that contains only specimens of their own kind. The aquarium should have a sandy bed and plants along the edges or many hiding places of stone and wood. It prefers hard, alkaline water. Keep at temperatures of 24° – 27°C (75° – 80°F).
Notes
The newly laid eggs are taken up in the mouth of the female and incubated while she attempts to gather the male's "egg spots," spots similar to the eggs on their anal fin. She keeps the eggs in her mouth until they hatch, in the typical manner of mouth-brooding species. As scientists study aquarium fish in more detail, so their scientific names may change. This is especially true of the family Cichlidae at the present moment.

Gnathonemus petersi

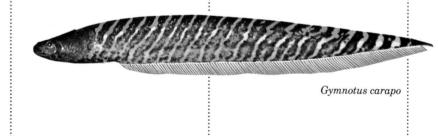

Gymnotus carapo

Gymnocorymbus ternetzi

Haplochromis burtoni

HELOSTOMA TEMMINCKI
Fam. Helostomatidae

Common name
Kissing gourami.
Dimensions
Up to 30 cm (12 in).
Distribution
Indochinese peninsula to Indonesia.
Food
In nature: mainly herbivorous. In aquarium: dried fish food, mixed with live food and vegetable matter to promote spawning. Feed 2 – 3 times a day.
General care
Easy to keep, adapting well to life with other species. It needs an aquarium of 75 liters (16½ gal) or more. The parents may eat the eggs, so they should be removed after they are laid. Initially the larvae require fine food until they can take infusorians, beaten egg yolk or dried fry fish food. After a week they can be given *Artemia* nauplii. Maintain temperatures at 24° – 30°C (75° – 86°F).
Notes
Wild specimens are silvery green; the pink variety available to aquarists is a domesticated race. There are no obvious sexual differences until the female is mature. The significance of the "kissing" is not wholly clear and seems not to be connected with copulation, given that it sometimes occurs among two fish of the same sex. No nest is built by spawning pairs. The eggs, laid at dusk in the course of the typical "kissing" activities of the adult fish, rise to the surface and hatch after about 24 hours.

HEMICHROMIS BIMACULATUS
Fam. Cichlidae

Common name
Jewel cichlid.
Dimensions
Up to 15 cm (6 in).
Distribution
Tropical Africa.
Food
In nature: carnivorous. In aquarium: live food or dried fish food mixed with insects, worms, fish and minced meat.
General care
An aggressive species. Keep alone in an aquarium provided with hiding places and constructed in such a way as to accommodate burrowing activity. No restrictions on water quality. Maintain temperatures at 22° – 28°C (72° – 82°F). Use a good external filter.
Notes
The colour is variable: sexually mature males are more reddish, but there are domesticated varieties that are entirely red. Eggs are laid on the substrate and the fish are noted for their parental care. The eggs hatch within 2 – 3 days; the larvae are transferred by mouth into a trench and tended by both parents until they measure 1 cm (½ in) or more. They start to swim and feed after 3 – 5 days.

HEMIGRAMMUS OCELLIFER
Fam. Characidae

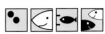

Common name
Beacon fish.
Dimensions
Up to 4 cm (1½ in).
Distribution
Amazon basin, Guyana.
Food
In nature: insects and other small organisms. In aquarium: any good quality dried fish food.
General care
A hardy species, easy to keep and often recommended to beginners. There are no restrictions on water quality but it should be slightly acid for breeding and not too calcareous. Use an aquarium of at least 30 liters (6½ gal) with small-leaved plants. Recommended temperatures are 22° – 27°C (72° – 80°F).
Notes
This is one of many characin species with iridescent spots on the caudal peduncle and on the eye. It can be found in specialized aquarium shops.

HYPHESSOBRYCON PULCHRIPINNIS
Fam. Characidae

Common name
Lemon tetra.
Dimensions
Up to 5 cm (2 in).
Distribution
Amazon basin.
Food
In nature: probably insects, small crustaceans and other tiny organisms. In aquarium: adapts to most types of dried fish food.
General care
As for *Hemigrammus ocellifer* but tends to be more shy and less demanding.
Notes
The male may be recognized by his slimmer profile and chiefly for the broad dark band that borders the anal fin. A placid species, it forms small shoals that prefer to settle in the lower half of the aquarium. Spawning can be encouraged by supplying a particularly varied diet for the females.

Hemichromis bimaculatus

Hyphessobrycon pulchripinnis

Helostoma temmincki

Hemigrammus ocellifer

TROPICAL FRESH WATERS • 49

HYPHESSOBRYCON SERPAE
Fam. Characidae

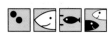

Common name
Serpae tetra.
Dimensions
Up to 5 cm (2 in).
Distribution
Amazon basin, Guyana.
Food
In nature: insects and other small organisms. In aquarium: thrives well on dried fish food.
General care
As for *Hemigrammus ocellifer*.
Notes
Fish of this species are short-lived (about 2 years) and succumb easily to dropsy. *H. serpae* is closely related to *H. callistus*, with which it is sometimes confused. However, the latter is less vividly coloured. In some varieties of *H. serpae* the typical rounded dorsal spot is absent. The eggs and newly hatched fry are almost colourless.

HYPOSTOMUS PLECOSTOMUS
Fam. Loricariidae

Dimensions
Up to 30 cm (12 in).
Distribution
Northern areas of South America, including Venezuela and Guyana.
Food
In nature: mainly herbivorous. In aquarium: feeds on surface algae but also needs an integrated diet of lettuce and fish food in a tank with plenty of plants.
General care
Thrives best in a large aquarium with hiding places and in company with other species. It is most active at dawn and dusk, and is a good tank cleaner. Keep at temperatures of 20° – 25°C (68° – 77°F). No restrictions on water quality.
Notes
On the basis of the little information available on breeding, it appears that the eggs change colour during embryo development. They hatch after 9 – 12 days and the male helps the young to crack open the egg shell. Also known as *Plecostomus plecostomus*.

ILYODON FURCIDENS
Fam. Goodeidae

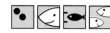

Dimensions
Up to 10 cm (4 in).
Distribution
West-central Mexico.
Food
In nature: omnivorous. In aquarium: feed with live food and dried fish food, including *Artemia* nauplii.
General care
This species prefers an aquarium of 60 liters (13 gal) or more, with surface plants and fairly dim lighting. The water should be slightly alkaline. Maintain temperatures at 24° – 27°C (75° – 80°F). Keep two males with between three and eight females in a tank reserved for this species.
Notes
In both sexes the body colour is fundamentally brownish and paler on the belly. The male is adorned with black spots and sometimes appears completely black. His dorsal and caudal fins have brown stripes and spots, with almost black margins. The females have bands across the rear part of the body. The male's anal fin is divided, whereas that of the female is rounded.

KRYPTOPTERUS BICIRRHIS
Fam. Siluridae

Common name
Glass catfish, ghost fish.
Dimensions
Up to 15 cm (6 in).
Distribution
Indochina and Indonesia.
Food
In aquarium: larvae of live insects, *Artemia* and other planktonic crustaceans; dried fish food.
General care
A tranquil, gregarious species, it is best to keep several together. No restrictions on water quality. Keep temperatures at 20° – 26°C (68° – 79°F).
Notes
This is a very transparent fish. The body is highly compressed with a tiny dorsal fin reduced to a single ray, the anal fin is very long and there is an adipose fin and a pair of mobile barbels on the upper jaw.

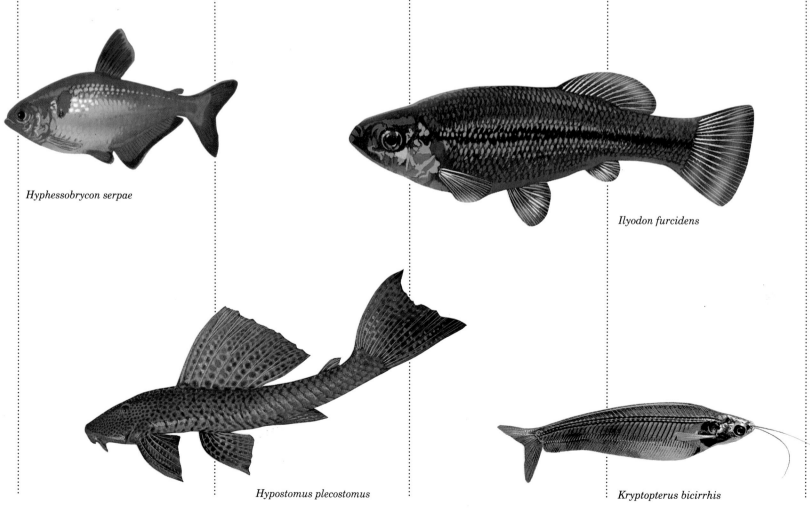

Hyphessobrycon serpae

Ilyodon furcidens

Hypostomus plecostomus

Kryptopterus bicirrhis

LABEO ERYTHRURUS
Fam. Cyprinidae

Common name
Red-finned shark.
Dimensions
Up to 12 cm (5 in).
Distribution
Thailand.
Food
In nature: omnivorous. In aquarium: standard diet including dried fish food containing vegetable matter.
General care
They need an aquarium of at least 125 liters (27½ gal) with softish, slightly acid water and plenty of vegetation.
The fish tolerate temperature fluctuations between 20° and 30°C (68° and 86°F). They tend to be aggressive with their kind and if several individuals are kept, they must be provided with hiding places. They are happy living in a community.
Notes
Many species of *Labeo* have become very popular with aquarists, some of whom consider *L. bicolor* to be the nicest of the group. This fish lives in the streams of Thailand, like *L. erythrurus*. Several *Labeo* species have now been bred in captivity. Unlike most of the Cyprinidae, this species lays only a few eggs (30 – 90), which are carried away from the male, who then ventilates them for the 2 – 3 days needed for hatching. The larvae begin to feed 2 – 3 days after they hatch.

LAMPROLOGUS BRICHARDI
Fam. Cichlidae

Dimensions
Up to 10 cm (4 in).
Distribution
Lake Tanganyika.
Food
In nature: small crustaceans and other invertebrates. In aquarium: preferably live food, mixed with frozen plankton to deepen colour; but fish food, flaked, frozen or dried, mixed with ground shrimps, worms, etc, is also suitable.
General care
A sociable species. A pair can live in an 80-liter (17½-gal) aquarium, but if more species are to be kept, the tank volume should be 250 liters (55 gal) or more. Cover the bed with gravel and rock shelters with plenty of hiding places. Plants are unnecessary. The water must be alkaline, hard and still. Keep the temperatures at 23° – 28°C (74° – 82°F). An efficient filtration system with a good biological filter is required. Keep tank well covered.
Notes
The female prepares the nest on the substrate and guards the newly laid eggs. Both parents care for the larvae which normally take *Artemia* nauplii, although at first they need infusorians or other forms of very small food.

LEPIDOSIREN PARADOXA
Fam. Lepidosirenidae

Common name
South American lungfish.
Dimensions
Up to 125 cm (50 in).
Distribution
Rivers and streams of Central and South America.
Food
In nature: benthic fish and invertebrates. In aquarium: takes small pieces of meat, fish, crustaceans and worms. It may be necessary to give live food, such as guppies, to initiate feeding.
General care
This species, one of the lungfish, is not easily available to aquarists; nevertheless, it is hardy and can withstand a wide range of conditions at temperatures of up to 20°C (68°F). Leave some air space above the aquarium to allow the fish to breathe at the surface.
Notes
In the spawning period, which occurs at the start of the rainy season, the male's ventral fins are modified, developing numerous branched vascular filaments. Some believe them to be supplementary breathing organs.

LEPORINUS FASCIATUS
Fam. Anostomidae

Dimensions
Up to 30 cm (12 in).
Distribution
Most of South America north of Rio de la Plata.
Food
In nature: omnivorous, its diet also including fruit, seeds and leaves. In aquarium: principally vegetable substances; supplement dried fish food with fresh vegetable matter.
General care
It needs a large aquarium with plenty of sturdy plants. No restrictions on water quality. Keep temperatures at 22° – 28°C (72° – 82°F).
Notes
This is a placid fish which lives happily with other species, even though its large size may induce it to become aggressive in a confined space. Because of its habit of leaping out of the water, the tank must be provided with a strong cover.

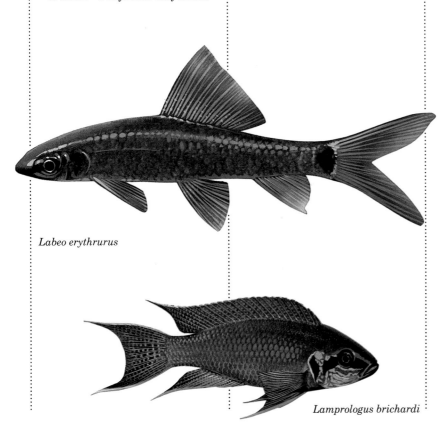

Labeo erythrurus

Lamprologus brichardi

Lepidosiren paradoxa

Leporinus fasciatus

TROPICAL FRESH WATERS • 51

LORICARIA FILAMENTOSA
Fam. Loricariidae

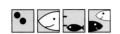

Common name
Whiptail catfish.
Dimensions
Up to 25 cm (10 in).
Distribution
Rio Magdalena, South America.
Food
In aquarium: vegetable matter mixed with normal dried fish food.
General care
This species is suitable for keeping with others and has been bred in captivity. It prefers softish, slightly acid water, at temperatures of 21° – 28°C (70°F – 82°F). Create hiding places and surfaces for growth of algae.
Notes
The male is distinguished from the female by small silky tufts on the head. When spawning, the pair choose a flat rock on which 100 – 200 amber eggs, each about 2 mm in diameter, are laid. They are attentively guarded by the male who ventilates them with his fins until they hatch after some 9 days. The fry should initially be raised in shallow water.

MACROPODUS OPERCULARIS
Fam. Belontiidae

Common name
Paradise fish.
Dimensions
Up to 10 cm (4 in).
Distribution
Warm temperate and tropical regions of southeast China, Korea, Vietnam and Taiwan.
Food
In nature: carnivorous, including insects and other invertebrates. In aquarium: accepts most forms of fish food, including worms, flatworms and crustaceans.
General care
It can survive in a small tank (10 liters [2 gal]) but it is better to use those of 40 liters (8¾ gal) or more. Although aggressive, it will live together with large, quiet species. The male builds an air-bubble nest. After egg laying, the female should be removed. The larvae are easily raised and when free-swimming will accept *Artemia* nauplii, infusorians, etc. Introduce vegetation, including floating plants. No restrictions on water quality; maintain temperatures of 15° – 24°C (59° – 75°F).
Notes
Although highly aggressive fish, they can be kept with other species. They have been bred in captivity. The sexes are difficult to distinguish.

MALAPTERURUS ELECTRICUS
Fam. Malapteruridae

Common name
Electric catfish.
Dimensions
Up to 1 m (3 ft).
Distribution
Central Africa, including Zaire and Nile basins.
Food
In nature: predatory. In aquarium: worms, strips of meat and fish.
General care
A hardy species but not compatible with other fish. Keep isolated in a suitably large aquarium at temperatures of 22° – 29°C (72° – 84°F). No restrictions on water quality.
Notes
This species is not normally kept in home aquaria. Its habits, and notably that of emitting electric shocks (which should not be tested), are such that there is little to enthuse the average aquarist once the novelty of owning this comparatively rare fish has worn off.

METYNNIS HYPSAUCHEN
Fam. Characidae

Common name
Silver dollar.
Dimensions
Up to 18 cm (7 in).
Distribution
Guyana, Amazonas and Paraguay.
Food
In nature: plants. In aquarium: fish food in pellet and flake form, mixed with lettuce and insects.
General care
Keep in a group in a large aquarium. The fish is not too demanding as to water quality but thrives best if it is not too calcareous and slightly acid, maintained at temperatures of 24° – 31°C (75° – 88°F). Use roots as shelter and plastic plants, if desired.
Notes
It breeds in a similar manner to other characins. After courtship, the two partners press against each other and, quivering, emit eggs and sperm. The eggs, 2 cm (¾ in) across, are not adhesive and hatch within 2 – 3 days. The larvae begin to swim and feed after 4 – 5 days and usually take live *Artemia* nauplii.

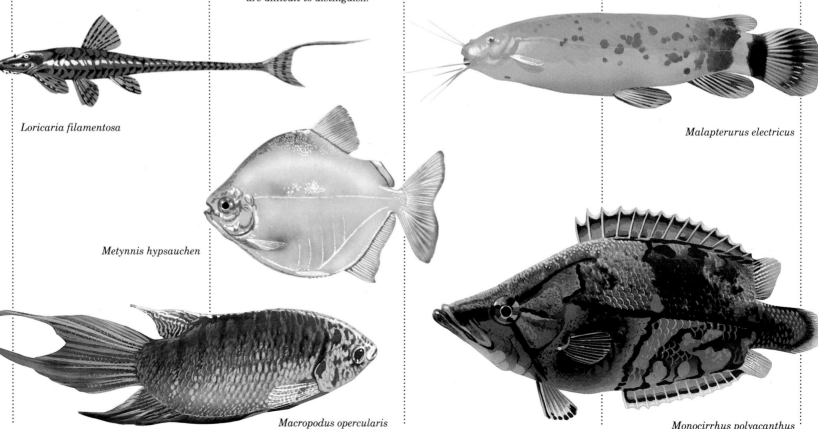

Loricaria filamentosa

Malapterurus electricus

Metynnis hypsauchen

Macropodus opercularis

Monocirrhus polyacanthus

MONOCIRRHUS POLYACANTHUS
Fam. Nandidae

Common name
Leaf fish.
Dimensions
Up to 10 cm (4 in).
Distribution
Basins of Amazon and Rio Negro, Guyana.
Food
In nature: tetras and other small fish. In aquarium: guppies and other small fish.
General care
Keep in a 75 liter (16½ gal) aquarium reserved for this species, with plenty of vegetation and softish, slightly acid water, at temperatures of 22° – 26°C (72° – 79°F).
Notes
It breeds in a similar manner to the cichlids. The male selects a zone on the lower surface of a leaf or stone, or, in the aquarium, of a flower container, and thoroughly cleans it. After a brief courtship, the female lays her eggs which stick to the chosen zone. Afterwards the male sprays them with his "milk" and guards them until they hatch, within 3 – 4 days; he also tends the free-swimming young for a further week.

MONODACTYLUS ARGENTEUS
Fam. Monodactylidae

Common name
Finger fish, mono.
Dimensions
Up to 25 cm (10 in).
Distribution
East coast of Africa to tropical western Pacific.
Food
In nature: omnivorous. In aquarium: adapts to a mixed diet of live and dried fish food.
General care
Very young specimens can be kept in fresh water but as they grow they need brackish water, while fully adult individuals must have sea water. Keep several specimens together in a large aquarium of 500 liters (110 gal) or more so that there is ample space for swimming. Recommended temperatures are 24° – 28°C (75° – 82°F).
Notes
The related *Monodactylus sebae*, originally from the western coasts of Africa, is distinguished by its body, which is higher than it is long because of the bigger fins, and by its vertical black stripes. Although uncommon in aquaria, it is nevertheless easy to breed once over the size of 10 cm (4 in). Each female lays 15,000 – 20,000 eggs which hatch in less than 24 hours.

NANNOSTOMUS HARRISONI
Fam. Lebiasinidae

Common name
Golden pencil fish.
Dimensions
Up to 6 cm (2½ in).
Distribution
Guyana, upper and middle Amazon.
Food
In nature: mainly small insects on the water surface but also feeds below the surface. In aquarium: small live insects such as fruit flies and ants; dried and frozen fish food.
General care
Prefers living in groups of six or more, in softish, slightly acid, peaty water, kept at temperatures of 23° – 26°C (74° – 79°F). Will adapt to small aquaria. Cover the bed with small-leaved plants.
Notes
Very common aquarium fish. Most are brightly coloured, active and conspicuous. They change colour each day, taking on stripes during the day and dark spots at night. Males of *N. harrisoni* normally have a white mark at the tip of the pectoral fins; females have a more rounded belly. The water in the breeding tank should be fairly soft and slightly acid, with numerous small-leaved plants. After the eggs are laid, the male ventilates them until they hatch within 2 days. The larvae start to swim after 4 days. Feed them with rotifers and infusorians, followed by *Artemia* nauplii.

NEOCERATODUS FORSTERI
Fam. Ceratodontidae

Common name
Australian lungfish.
Dimensions
Up to 180 cm (72 in).
Distribution
Queensland, Australia.
Food
In nature: fish, amphibians, invertebrates and some vegetable matter, probably ingested because it contains tiny organisms.
Notes
Neoceratodus, because of its characteristics, is considered to be the most primitive of the three present-day genera of lungfish, the other two being *Lepidosiren* and *Protopterus*, all belonging to the order Dipnoi. The Australian species, in fact, has a monolobate lung and cannot live out of the water in pockets of dry mud. It is rarely seen except in public aquaria and is a protected species.

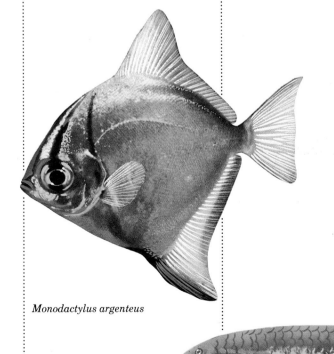

Monodactylus argenteus

Nannostomus harrisoni

Neoceratodus forsteri

TROPICAL FRESH WATERS • 53

NOTOPTERUS CHITALA
Fam. Notopteridae

Common name
Indian knife fish.
Dimensions
Up to 90 cm (36 in).
Distribution
Tropical fresh waters of Burma, India and Thailand.
Food
In nature: a predator that feeds on small fish and invertebrates. In aquarium: normally accepts pieces of fish and meat, but small fish are sometimes necessary as first food.
General care
Being a large species, only young specimens are suitable for bigger home aquaria. Water quality does not pose serious problems but ideally it should be fairly soft, enriched with humus and maintained at temperatures of 24° – 28°C (75° – 82°F).
Notes
This species is aggressive and should be kept isolated or perhaps with other large fish of different species. A smaller species from Africa, *Xemomystus nigri*, is sometimes obtainable from specialist outlets; it is less beautiful but better suited to home aquaria.

ORYZIAS LATIPES
Fam. Cyprinodontidae

Common name
Japanese medaka, rice fish.
Dimensions
Up to 4 cm (1½ in).
Distribution
Japan.
Food
In aquarium: prefers small live food but also adapts to suitable dried fish food.
General care
Lives well in groups and with other quiet species. Keep in an aquarium of 12 liters (2½ gal) or more, with floating plants and slightly alkaline water at temperatures of 20° – 25°C (68° – 77°F).
Notes
This is one of the many species of *Oryzias* present in Southeast Asia, in various habitats such as rice paddies, ditches and mountain streams. The females are courted vigorously by the males. Fertilization is internal. The eggs remain hanging to the female's body in bunches and then become attached to floating vegetation.

OSPHRONEMUS GORAMY
Fam. Osphronemidae

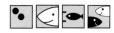

Common name
Giant gourami.
Dimensions
Up to 60 cm (24 in).
Distribution
Borneo, Sumatra, Java. It has been imported into other areas of Southeast Asia and into Australia, mainly because its flesh makes good eating.
Food
In nature: omnivorous. In aquarium: accepts fresh animal and plant foods as well as dried and freeze-dried fish food.
General care
In a large aquarium the species grows so rapidly that it may become a nuisance. Young specimens may be kept in a community tank with other placid species. There are no restrictions on water quality, but the fish must be given plenty of room for swimming and some floating plants among which the male can build his bubble nest and nibble at fragments for additional sustenance. The eggs are carefully tended by the male until they hatch within 24 – 36 hours. The fry are free-swimming after a few days and can be fed initially with rotifers and infusorians.
Notes
The male's dorsal and anal fins are elongated and pointed at the rear.

OSTEOGLOSSUM BICIRRHOSUM
Fam. Osteoglossidae

Common name
Arowana.
Dimensions
Up to 1 m (3 ft).
Distribution
Guyana and Amazon basin.
Food
In nature: this species swims just beneath the water surface feeding on a wide variety of live organisms such as insects (beetles), spiders, crustaceans, gastropods and, to a lesser extent, fish and suspended detritus. In aquarium: a variety of items, including small fish, *Artemia*, *Daphnia*, insect larvae, meat worms and dried fish food.
General care
Select a tank to suit the size of the fish but preferably not under 400 liters (88 gal); transfer into a larger aquarium as the fish grows. The Osteoglossidae prefer water that is not too calcareous, enriched with humus, at temperatures of 23° – 27°C (74° – 80°F). The aquarium must be well sealed as this fish is a remarkable jumper.
Notes
A rare aquarium species and very handsome with its big eyes, silvery scales and sprouting barbels on the lower jaw. It is particularly attractive in a tank filled with plants. It is an easy fish to keep and may live several years in captivity. It is a mouth-brooder: after the eggs are laid, one of the parents collects the fertilized eggs and holds them in the mouth until they hatch. The fish are difficult to breed or to raise in captivity.

Notopterus chitala

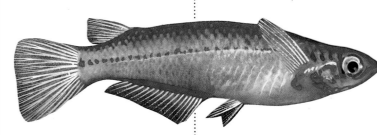

Oryzias latipes

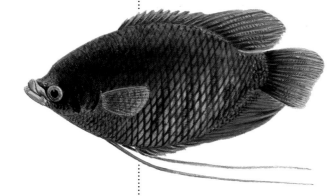

Osphronemus goramy

Osteoglossum bicirrhosum

PANTODON BUCHHOLZI
Fam. Pantodontidae

Common name
Butterfly fish.
Dimensions
Up to 10 cm (4 in).
Distribution
Calm waters and pools associated with the rivers of tropical West Africa.
Food
In nature: a wide variety of surface organisms including fish, insects and invertebrates. In aquarium: a broad range of items including meat worms, grubs and dried fish food.
General care
Small specimens can be kept in a 75-liter (16½-gal) aquarium but 300-liter (66-gal) tanks are better. The tank must be well sealed because the fish is a good jumper. A tranquil species, it lives at the surface, providing an interesting contrast to other species kept in the tank. It is advisable to use plants with floating leaves. Maintain the water at temperatures of 25° – 30°C (77° – 86°F).
Notes
The butterfly fish has been successfully bred in captivity. There is a long courtship and the male grasps the female as she lays the eggs. Fertilization may be internal. The newly laid eggs float on the surface and hatch within about 2 days. The larvae are very small and only take tiny food items.

PAPILIOCHROMIS RAMIREZI
Fam. Cichlidae

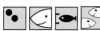

Common name
Ramirez' dwarf cichlid.
Dimensions
Up to 5 cm (2 in).
Distribution
Venezuela.
Food
In nature: aquatic insects and crustaceans. In aquarium: live food mixed with dried fish food.
General care
A tranquil species. Use an aquarium with vegetation and hiding places. It prefers only slightly calcareous, acid water. Maintain temperatures of about 25°C (77°F).
Notes
It is less territorial than other cichlids and because of its shy, quiet nature, needs to be kept on its own. The fish is highly susceptible to dropsy, tuberculosis and *Icthyosporidium* infections. It is sometimes referred to as *Microgeophagus ramirezi*.

PARACHEIRODON INNESI
Fam. Characidae

Common name
Neon tetra.
Dimensions
Up to 4 cm (1½ in).
Distribution
Upper Amazon basin.
Food
In nature: prefers insects and their larvae. In aquarium: frozen, dried and other forms of fish food, together with live insects.
General care
Can be kept in small aquaria, living best in groups of six or more individuals. The water should be not too calcareous and slightly acid, obtained by peat filtration. Introduce small-leaved plants and provide moderate lighting. Temperatures for breeding should be 21° – 23°C (70° – 74°F).
Notes
Neons live in pools and quiet jungle streams. Similar in appearance to *Cheirodon axelrodi*, they are rather smaller and the front of the body is less red.

PELMATOCHROMIS KRIBENSIS
Fam. Cichlidae

Common name
Dwarf rainbow cichlid, krib.
Dimensions
Up to 10 cm (4 in).
Distribution
West Africa, delta of River Niger.
Food
In aquarium: prefers live food such as worms, insects and crustaceans. It should adapt to a diet of good dried fish food, mixed occasionally with live items.
General care
A fairly tranquil species which can be kept with others in an aquarium of 55 liters (12 gal) or more, with plants. Introduce stones and roots to make hiding places.
Notes
The fish lays eggs in caves, both parents caring for the eggs and larvae. Aquarists often introduce pots laid on their sides, for use as nest sites. The male is bigger than the female and has a more pointed tail, with 2 to 3 spots on the upper lobe. The pH of the water appears to influence the sex of the larvae; very acid water increases the percentage of females while neutral water encourages the birth of males. It is also known as *Pelvicachromis pulcher*.

Papiliochromis ramirezi

Paracheirodon innesi

Pantodon buchholzi

Pelmatochromis kribensis

PHENACOGRAMMUS INTERRUPTUS Fam. Characidae

Common name
Zaire or congo tetra.
Dimensions
Up to 8 cm (3 in).
Distribution
Zaire basin.
Food
In nature: insects and other small organisms. In aquarium: frozen, dried or otherwise prepared fish food, insects, *Artemia*.
General care
Easy to keep. Lives best in uncrowded aquaria of 75 liters (16½ gal) or more, with vegetation. Water should not be too calcareous, slightly acid and maintained at temperatures of 25° – 27°C (77° – 80°F).
Notes
Perhaps the most beautiful of African tetras, it has a lobate caudal fin and delicate colours. The long rays of the male's dorsal and caudal fins grow better if the tank water is softish and slightly acid, and the tank spacious enough for swimming, with a fairly dark bed. Breeding is stimulated by sunlight, the eggs hatching within some 6 days. The species is also classified as *Micralestes interruptus* and can be placed in the family Alestidae.

POECILIA RETICULATA Fam. Poeciliidae

Common name
Guppy, millions fish.
Dimensions
Up to 5 cm (2 in); males smaller.
Distribution
Northern areas of South America, West Indies.
Food
In nature: omnivorous, including insect larvae and crustaceans. In aquarium: dried fish food, insects, worms, *Artemia*.
General care
Easy to keep and breed. Normally a trap is used for breeding. Maintain temperatures at 20° – 28°C (68° – 82°F).
Notes
This is one of the most popular aquarium fish because of its coloration and ease of reproduction. There are specialized guppy societies in many countries and aquarists have the choice of many varieties that are distinguished basically by shape of tail (fan, lyre, double lyre, sail, rounded, etc.). The species has been known since 1860 when several specimens from Trinidad were donated to the British Museum by Rev. R. J. L. Guppy, whose name has been appropriately commemorated.

POECILIA SPHENOPS Fam. Poeciliidae

Common name
Short-finned molly.
Dimensions
Up to 12 cm (5 in); males smaller.
Distribution
From Texas to Mexico and Central America to Colombia and Venezuela.
Food
In aquarium: dried fish food containing vegetable matter.
General care
Prefers slightly brackish water. Adapts well to living with other species. There are many varieties.
Notes
Males of *P. sphenops* were formerly used to obtain the celebrated "black molly" variety. In the aquarium excessive movement of the fins may indicate the presence of a bacterial infection. The species is sensitive to fluctuations of temperature which are liable to bring about inflammation of the swim bladder.

POECILIA VELIFERA Fam. Poeciliidae

Common name
Sailfin molly.
Dimensions
Up to 8 cm (3 in).
Distribution
Coastal regions of South Carolina to northeast Mexico. Introduced into other areas, including California.
Food
In nature: detritus and algae. In aquarium: normal dried aquarium diet, mixed with vegetable matter.
General care
Keep in pairs or small groups in a large tank of 75 liters (16½ gal) or more, with plenty of vegetation. No restrictions on water quality; maintain temperatures at 24° – 28°C (75° – 82°F). The species is easy to breed, gestation lasting 8 – 10 weeks. Provide a tank with plenty of plants. The 20 – 80 young immediately accept *Artemia* nauplii or other similar forms of live food.
Notes
The main characteristic of this species is the large dorsal fin, which may cause it to be confused with *P. latipinna*. In fact, the latter has fewer rays (14 instead of 18 – 19) and light spots at the base of the elongated rather than rounded fin.

Phenacogrammus interruptus

Poecilia reticulata

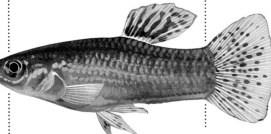

Poecilia sphenops

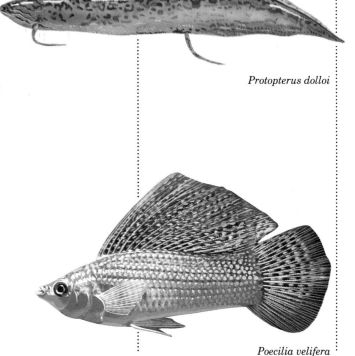

Protopterus dolloi

Poecilia velifera

PROTOPTERUS DOLLOI
Fam. Protopteridae

Common name
African lungfish.
Dimensions
Up to 85 cm (34 in).
Distribution
Zaire basin.
Food
In nature: slow-swimming fish and invertebrates. In aquarium: lean meat and fish. If necessary, try guppies and small gastropods as first food.
General care
Small specimens of this lungfish can be kept in a domestic aquarium and they are sometimes available as tropical aquarium fish. The species shows aggressiveness towards others and is best kept isolated. It prefers temperatures of 20° – 30°C (68° – 86°F) but will survive at higher levels. Leave a space between the water surface and the cover.
Notes
The fins of this lungfish are thin and very mobile: when it moves slowly along the bottom, it gives the impression of walking. The fish is sought by the natives for its appetizing flesh, particularly during the dry season when its body is full of fatty substances.

PSEUDOTROPHEUS ZEBRA
Fam. Cichlidae

Common name
Zebra Malawi cichlid.
Dimensions
Up to 16 cm (6¼ in).
Distribution
Lake Malawi.
Food
In nature: strips algae from rocks. In aquarium: dried fish food rich in vegetable matter mixed with live food.
General care
Keep one male with two or more females in a tank of 400 liters (88 gal) or more. Prepare rock burrows or pottery containers as refuge for the female. The water should preferably be slightly alkaline and hard, with temperatures of 28° – 30°C (82° – 86°F).
Notes
The female gathers her eggs immediately after laying them, and then nibbles the belly of the male to induce him to emit sperm. Fertilization occurs in the female's mouth and she broods the eggs there until they hatch (within about 20 days). She then continues to care for the larvae. *P. zebra* is a polymorphic species with various natural forms and colours, including red and orange, and additional varieties produced in the aquarium. Other well-known species are *P. auratus* and *P. elongatus*.

PTEROPHYLLUM SCALARE
Fam. Cichlidae

Common name
Angelfish.
Dimensions
Up to 15 cm (6 in).
Distribution
Amazon basin.
Food
In nature: small live organisms, including crustaceans and insect larvae. In aquarium: crustaceans, insects, worms and dried fish food.
General care
Use a large, deep aquarium with vegetation around the walls (*Echinodorus*, etc.). The water should range from neutral to slightly acid and temperatures from 25° to 27°C (77° to 80°F). Keep various angelfish together and allow them to form pairs freely. The fish lay the eggs on slates or flat stones; the eggs can be removed in order to raise the young separately, but also left with the parents. Wean the larvae with boiled, chopped egg yolk, followed by *Artemia* nauplii. The fish adapt to living with other tranquil species.
Notes
There are three species of *Pterophyllum*: *P. scalare*, *P. altum* and *P. dumerilii*. In addition, numerous varieties have been produced by artificial selection, including albino individuals.

RASBORA HETEROMORPHA
Fam. Cyprinidae

Common name
Harlequin fish, red rasbora.
Dimensions
Up to 5 cm (2 in).
Distribution
Southeast Asia, Sumatra.
Food
In nature: a variety of small aquatic organisms living at the water surface. In aquarium: they adapt well to dried fish food but live foods (insects, crustaceans and worms) stimulate spawning.
General care
Keep more than one together. They can be bred in small aquaria. The water should be acid and not too calcareous. To encourage spawning, maintain temperatures of 23° – 27°C (74° – 80°F). The tank should contain large-leaved plants and be roomy enough to allow the formation of groups. The eggs are attached to the lower surface of the leaves and hatch within 24 hours, when the adults should be removed.
Notes
An ideal aquarium fish. It is brightly coloured, sturdy and active, yet docile. It lives successfully with others of its kind and with other tranquil species. For a small fish, it is very long-lived, reaching five years or more in captivity.

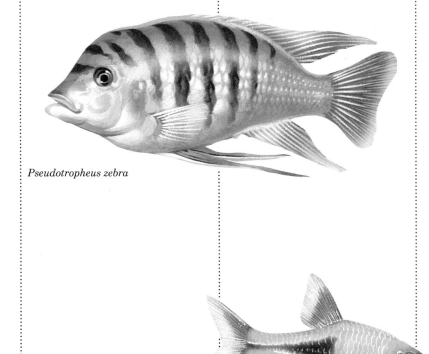

Pseudotropheus zebra

Rasbora heteromorpha

Pterophyllum scalare

TROPICAL FRESH WATERS • 57

RASBORA MACULATA
Fam. Cyprinidae

Common name
Spotted rasbora.
Dimensions
Up to 2.5 cm (1 in).
Distribution
Southeast Asia.
Food
In aquarium: readily accepts dried or freeze-dried fish foods; mix with fresh food (worms and small crustaceans).
General care
The species should be kept in small shoals, in tanks with dark bottoms and plenty of vegetation. The water should be soft and slightly acid, filtered with peat.
Notes
This is the smallest *Rasbora* species. The male is smaller and slimmer, coloured cherry red. The female is yellowish, with a typically rounded belly outline. At time of spawning, pairs should be separated for a couple of weeks and fed on a very varied diet, in larger amounts than usual. Spawning occurs within a few days of the pair formation. The adults should be separated after the laying of the eggs, which hatch within 24 – 30 hours.

RASBORA TRILINEATA
Fam. Cyprinidae

Common name
Scissors-tail rasbora.
Dimensions
Up to 6 cm (2½ in).
Distribution
Southeast Asia, Sumatra.
Food
As for *R. heteromorpha*.
General care
The scissors-tail name derives from the movement of the lobes of the caudal fin, which open and close like scissors. Females are distinguished by their more rounded abdomen. The water in the breeding tank should be softer than that in which the adults are normally raised (10˚ – 12˚GH). The eggs, laid freely in the water, hatch in under 30 hours, and as for other rasboras, the adults should be removed from the tank where they have paired.

SAROTHERODON MARIAE
Fam. Cichlidae

Common name
Tilapia.
Dimensions
Up to 15 cm (6 in).
Distribution
West Africa.
Food
In nature: principally herbivorous. In aquarium: vegetable matter mixed with live food; dried fish food.
General care
Use a large aquarium with rocks and wood as shelters, built strongly enough to withstand burrowing activity; create caves with flat stones on top. There are no restrictions on water quality, but ideally it should be slightly acid. Maintain temperatures at 24˚ – 27˚C (75˚ – 80˚F).
Notes
This species lays eggs in the substrate; the female attaches the 200 – 400 eggs to the underside of a stone. After fertilization, she guards the eggs while the male watches over the territory. After 2 days the female transfers the eggs to a suitably prepared trench where they hatch within a few hours. The young start to swim after absorbing the vitelline sac, 5 – 6 days later, and can be fed on *Artemia* nauplii.

SCATOPHAGUS ARGUS
Fam. Scatophagidae

Common name
Scat, argus fish.
Dimensions
Up to 30 cm (12 in); smaller in captivity.
Distribution
Southeast Asia.
Food
In nature: omnivorous. In aquarium: dried fish food containing vegetable matter, mixed with live food.
General care
Can be kept with one or more species. Keep several specimens together in the largest possible aquarium. Cover the bed with sand and a few stones for decoration, without plants. Install an undergravel filter and a supplementary external filter because the fish excretes copiously. Brackish water is advisable; use about 60 g of sea salt to every 10 liters (2 gal) of fresh water. Maintain temperatures at 20˚ – 28˚C (68˚ – 82˚F).
Notes
They are usually gregarious fish and should not be isolated. They are somewhat nervous, however, so their tank must be placed in a quiet corner of the room.

Rasbora maculata

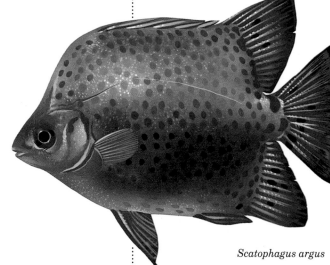

Scatophagus argus

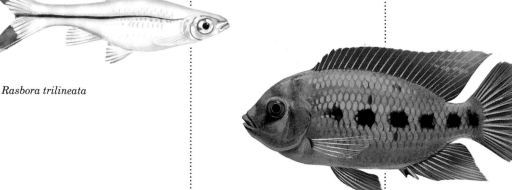

Rasbora trilineata

Sarotherodon mariae

SERRASALMUS NATTERERI
Fam. Characidae

Common name
Red piranha.
Dimensions
Up to 30 cm (12 in).
Distribution
Amazon and Orinoco basins.
Food
In nature: carnivorous; the sharp teeth enable the fish to feed on large prey, mainly fish but also other vertebrates such as mammals, or insects and invertebrates. In aquarium: pieces of fish mixed with meat.
General care
A large tank is essential. No restrictions on water quality. Keep temperatures at 21° – 27°C (70° – 80°F). Because it is so aggressive, it is best to start with a group of young individuals and raise them together in order to reduce fights.
Notes
There are some 15 species of piranha in the genus *Serrasalmus*; many are similar in appearance and hard to tell apart. *S. nattereri* is the one most commonly imported but it may be confused with other species.
S. rhombeus is very familiar to aquarists and has incorrectly been named *S. niger*. This group is continually subject to revisions, even to the extent of putting these fish in a separate family, the Serrasalmidae.

SYMPHYSODON DISCUS
Fam. Cichlidae

Common name
Discus.
Dimensions
Up to 15 cm (6 in).
Distribution
Amazon basin.
Food
In nature: omnivorous. In aquarium: live food, dried fish food containing vegetable matter, ox heart.
General care
This is a delicate species which needs a large aquarium with plenty of vegetation and water that is not too calcareous and slightly acid. A tank of their own is recommended but it is possible to accommodate them with other tranquil species, such as angelfish. Paint the base and sides of the aquarium black to reduce agitation, and use a branch or stump to create a shelter in the center of the tank. It prefers temperatures of 25°C (77°F) or slightly more.
Notes
A unique characteristic of *S. discus* is the manner in which the young are raised by feeding on mucus secreted by the skin of the parents.

SYNODONTIS ALBERTI
Fam. Mochokidae

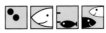

Dimensions
Up to 20 cm (8 in).
Distribution
Zaire basin.
Food
In nature: algae and other vegetable matter, insects and crustaceans. In aquarium: normal aquarium diet including dried foods.
General care
This is an ideal fish to keep with other species and in addition it rids the tank of unwanted algae. Keep a few individuals together at temperatures of 22° – 27°C (72° – 80°F).
Notes
The species has barbels on the upper jaw that are longer than those of other species belonging to the same genus. The lower barbels, on the other hand are branched, resembling the roots of aquatic plants.

SYNODONTIS NIGRIVENTRIS
Fam. Mochokidae

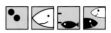

Common name
Upside-down catfish.
Dimensions
Up to 10 cm (4 in).
Distribution
Central Africa.
Food and general care
As for *S. alberti*.
Notes
This species has a dark abdomen and camouflage coloration when resting belly upwards, resembling the underside of a branch or leaf. It has been bred in captivity.

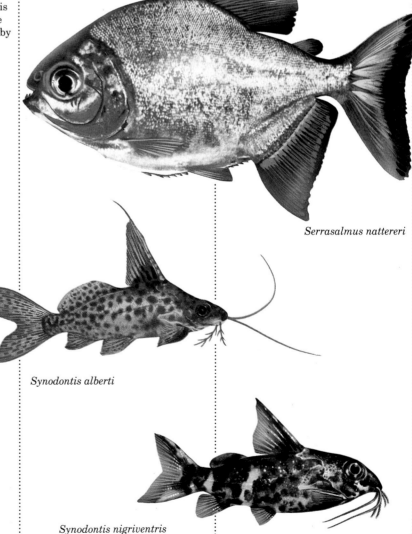

Serrasalmus nattereri

Synodontis alberti

Synodontis nigriventris

Symphysodon discus

TROPICAL FRESH WATERS

TETRAODON FLUVIATILIS
Fam. Tetraodontidae

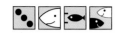

Common name
Green puffer fish.
Dimensions
Up to 18 cm (7 in).
Distribution
Southeast Asia.
Food
Omnivorous, especially live food.
General care
Needs neutral or slightly basic, hard water; for breeding add a small quantity of salt. The water temperature should be 22° – 26°C (72° – 79°F). Provide the tank with rock clefts and hiding places, plus fairly tough vegetation.
Notes
The young are peaceful but adults become aggressive. As they fill themselves with air or water, they puff up like balloons, with spines erect.

TETRAODON PALEMBANGENSIS
Fam. Tetradontidae

Dimensions
Up to 20 cm (8 in).
Distribution
Southeast Asia.
Food
In nature: carnivorous. In aquarium: thrives best if the diet includes gastropods; mix with worms, crustaceans and insects.
General care
Keep as single species in brackish water (about 20% of normal sea water). Use a conventional filter and provide a sandy bed with a few flat stones on which the eggs can be laid. The male guards and fans the eggs until they hatch, normally for 6 – 8 days. The fry are small and difficult to rear. Try to feed them with rotifers (e.g. *Brachionus*), followed by *Artemia* nauplii. Maintain water at temperatures of 23° – 28°C (74° – 82°F).

TOXOTES JACULATOR
Fam. Toxotidae

Common name
Archerfish.
Dimensions
Up to 24 cm (9½ in).
Distribution
From coasts of Red Sea, India and Sri Lanka to Australia and Melanesia.
Food
In nature: insects. In aquarium: insects, worms, chopped meat and fish.
General care
Use a tall aquarium of 125 liters (27½ gal) or more, filling it partly with fresh, sea or brackish water, and with adequate hiding places. Ground plants can be provided. Maintain temperatures of 22° – 28°C (72° – 82°F). The archerfish captures insects on plants above water level. The species is very difficult to breed in captivity.

TRICHOGASTER LEERI
Fam. Belontiidae

Common name
Pearl gourami.
Dimensions
Up to 11 cm (4½ in).
Distribution
Southeast Asia from Thailand to Borneo and Sumatra.
Food
In nature: carnivorous. In aquarium: worms, insects and chopped shrimp; dried fish food.
General care
Use an aquarium of 40 liters (8¾ gal) or more, with many plants, including floating species. Maintain temperatures at 23° – 30°C (74° – 86°F).
Notes
The abdomen of both sexes, when ready to spawn, may turn bright orange – red, though that of the female is much less vivid. The manner of breeding is as described for the family (see page 141). A quiet species, it is easy to keep and happy living in a community, although ideally it should have a tank on its own. Feed with live food to create ideal spawning conditions. The male is distinguished by his elongated dorsal fin.

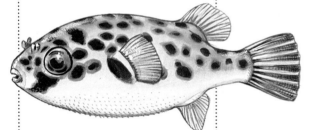

Tetraodon fluviatilis

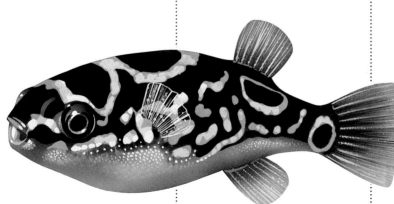

Tetraodon palembangensis

Toxotes jaculator

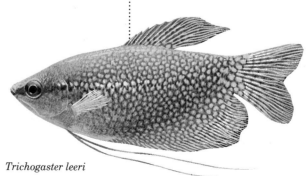

Trichogaster leeri

60 • TROPICAL FRESH WATERS

TRICHOGASTER TRICHOPTERUS
Fam. Belontiidae

Common name
Three-spot gourami.
Dimensions
Up to 13 cm (5¼ in).
Distribution
Southeast Asia from Thailand to Sunda Islands.
Food
In nature: insects, crustaceans, worms and other small invertebrates. In aquarium: dried fish food mixed with live food such as mosquito larvae, worms and *Daphnia*.
General care
Can be kept together with other species and in a group of its own kind, but in the latter case it may turn aggressive. Use an aquarium, with vegetation, of 40 liters (8¾ gal) or more. There are no restrictions on water quality but maintain temperatures of around 24° – 29°C (75 – 84°F).
Notes
The male's dorsal fin is higher and more pointed than that of the female. The common name of the species is derived from the two dark spots on the flanks, one in the center and one on the caudal peduncle, plus the black eye. Well stocked suppliers may offer the subspecies *T. trichopterus sumatranus*, the blue or opaline gourami, the basic colour being blue, offset by mother-of-pearl spots that adorn the fins.

TROPHEUS MOOREI
Fam. Cichlidae

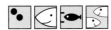

Dimensions
Up to 12 cm (5 in).
Distribution
East Africa: Lake Tanganyika.
Food
Tends to be omnivorous. Takes dried and freeze-dried foods, mixed with worms, crustaceans, insects and plant matter.
General care
Prefers hard water and tanks with rocky beds and floating plants that cut down light incidence. Plants that root in the bottom are unsuitable. Water temperatures should be 22° – 25°C (72° – 77°F).
Notes
This is a robust and aggressive African cichlid which is best kept in a small group in its own tank. It is a mouth-brooder, the female gathering the eggs into her mouth as soon as they are laid. Fertilization occurs directly inside her mouth, where the eggs remain for some 4 weeks before hatching. The fry swim out of the mother's mouth when they are about 1 mm long and already able to feed on small crustaceans and worms.

XIPHOPHORUS HELLERI
Fam. Poeciliidae

Common name
Swordtail.
Dimensions
Up to 12 cm (5 in); males slightly smaller.
Distribution
Central America.
Food
In nature: omnivorous. In aquarium: dried fish food, worms, insects, crustaceans and vegetable matter.
General care
A tranquil, active species, much favoured by discerning aquarists. Keep in a tank with vegetation, preferably in soft, slightly alkaline water at temperatures of 21° – 26°C (70° – 79°F). Will live happily with other placid species. Easy to breed. Gestation lasts 4 – 6 weeks. The young immediately accept *Artemia* nauplii and, after the second week, a more varied diet.
Notes
This is one of the most common and popular of the Poeciliidae. Individuals of this species do not take well to crowded tanks or to smaller fish which hinder their full development. The species is remarkably adaptable, so that in nature there exists a striking variety of subspecies, races and forms, giving aquarists the opportunity to selectively breed specimens of many different colours (red, green, variegated) and tail shapes.

XIPHOPHORUS MACULATUS
Fam. Poeciliidae

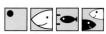

Common name
Platy.
Dimensions
Up to 6 cm (2½ in); males smaller.
Distribution
Central America, including southern Mexico, Honduras and Guatemala.
Food
In nature: omnivorous. In aquarium: dried fish food containing vegetable matter, insects, etc.
General care
Prefers slightly alkaline water in a well-planted aquarium. Gestation lasts about 4 weeks and 10 – 15 young are born, normally accepting *Artemia* nauplii.
Notes
The common name familiar to aquarists is an abbreviation of the scientific name *Platypoecilus maculatus* by which it was previously known. Thanks to crosses and selective breeding, hybrids have been created with characteristics that are now standardized. Among these varieties are the red platy, the golden platy, the black platy and the red crescent platy.

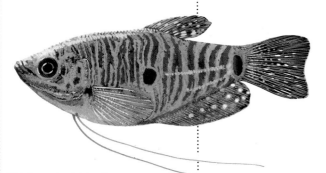

Trichogaster trichopterus

Xiphophorus helleri

Tropheus moorei

Xiphophorus maculatus

CARASSIUS AURATUS
Fam. Cyprinidae

Common name
Goldfish.
Dimensions
Up to 25 cm (10 in).
Distribution
Eastern Asia; widely distributed elsewhere.
Food
In nature: omnivorous, insects, molluscs, crustaceans, worms and aquatic vegetation. In aquarium: adapts to a broad range of diets; as a rule, dried foods are most convenient.
General care
These are easy fish to keep, either in small bowls or elaborate aquaria. The common varieties do not need heating or aeration. It is useful to put a few plants, such as *Elodea* (waterweed) into the tank. They are sturdy temperate fish, some of them ideal for pools and ponds, in which case they will tolerate temperatures of 0° – 38°C (32° – 100°F).
Notes
Sometimes confused with the carp, the goldfish may be identified by the absence of barbels. Thanks to artificial selection, a large number of standard varieties have been created.

COTTUS GOBIO
Fam. Cottidae

Common name
Miller's thumb.
Dimensions
Up to 17 cm (6½ in).
Distribution
Europe and Asia.
Food
In nature: carnivorous; aquatic insects and benthic invertebrates. In aquarium: adapts to most types of fish food but prefers live food.
General care
Needs cool water. Keep in an aquarium of 75 liters (16½ gal) or more, with a bed of pebbles and stones for shelter. It is best to have one male to several females. Eggs are laid beneath stones and the male guards the nest. In nature, the miller's thumb prefers clear, slow-flowing streams with stony bottoms. In the daytime it hides among the pebbles, feeding most actively at dawn and dusk.
Notes
This is a territorial species, so only limited numbers of males can be kept in the same tank, although a bed with plenty of crevices makes it possible to accommodate more individuals.

CYPRINUS CARPIO
Fam. Cyprinidae

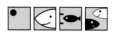

Common name
Koi carp.
Dimensions
Up to 100 cm (40 in).
Distribution
Not found in nature, although the wild form is.
Food
In captivity they are fed with pellets to which may be added carotenoid pigments to develop their colours.
General care
They are normally raised in shallow ponds with a base of cement or rock. Plants such as water lilies may be introduced, provided they are in containers. Place large stones on the bottom to prevent the fish burrowing under the plants. The species tolerates temperatures of 0° – 38°C (32° – 100°F).
Notes
Over the last century many marvellous varieties of Koi carp have been created by artificial selection, particularly by the Japanese. Some selected specimens fetch extremely high prices while more common ones are available cheaply. Unlike those species found in natural surroundings, which are difficult to discern in lakes and ponds, Koi are highly attractive either seen from above or through the glass of an aquarium.

GAMBUSIA AFFINIS
Fam. Poeciliidae

Common name
Spotted gambusia, western mosquito fish.
Dimensions
Females up to 6 cm (2½ in); males smaller, at most 3.5 cm (1½ in).
Distribution
Southeastern United States, Texas and Mexico. Introduced around the world.
General care
They are aggressive and bite the fins of other species so it is best to keep them on their own. It is a robust species, tolerating temperatures of 10° – 27°C (50° – 80°F). Females give birth to 1 – 130 young after a gestation of 3 – 4 weeks. The adults are cannibals.
Notes
Two species of *Gambusia* have been introduced to many countries to help wipe out malaria-transmitting mosquitoes. They are *G. affinis* and *G. holbrooki*.

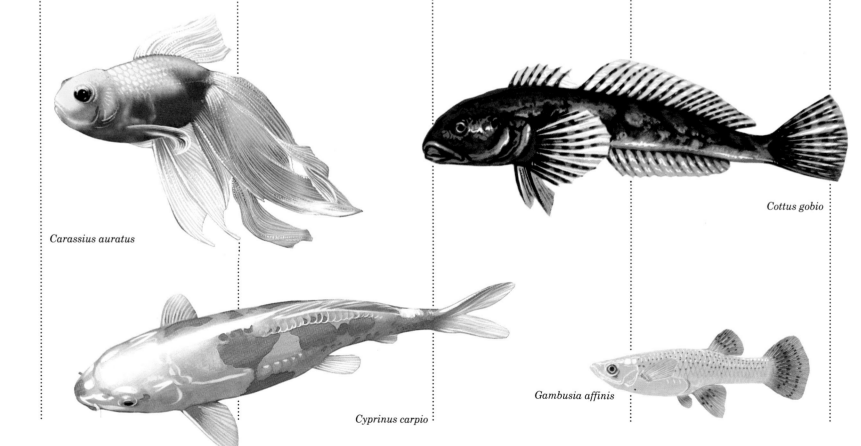

Carassius auratus

Cottus gobio

Cyprinus carpio

Gambusia affinis

GASTEROSTEUS ACULEATUS
Fam. Gasterosteidae

Common name
Three-spined stickleback.
Dimensions
Up to 8 cm (3 in).
Distribution
Widely found in Europe and across Russia to Japan and Korea; also along both coasts of North America.
Food
In nature: freshwater populations feed on benthic organisms and creatures living on plants, whereas migratory populations feed mainly on pelagic organisms (zooplankton, etc.). In aquarium: they adapt to almost all kinds of fish food.
General care
Keep at room temperature in water similar to that in which they live naturally. Provide a sandy bed and plenty of vegetation for building nests; use external or undergravel filters. It breeds freely in captivity if the tank is sufficiently large and there is adequate nest-building materials.
Notes
This is a species from temperate zones, found in fresh, brackish or sea water. Although not particularly spectacular in appearance, it is an interesting aquarium species because of its reproductive behaviour. Sticklebacks normally spawn in late spring. The male digs a small trench and builds a tunnel-shaped nest with a heap of vegetable matter. He may court and mate with several females but guards the eggs and the fry alone.

LEPOMIS GIBBOSUS
Fam. Centrarchidae

Common name
Common sunfish.
Dimensions
Up to 20 cm (8 in) but not often longer than about 15 cm (6 in).
Distribution
Originally from eastern North America, this fish is today acclimated in various parts of Europe where it has often become a pest inasmuch as it destroys the eggs and fry of more valuable fish species.
Food
In aquarium: aquatic insects and their larvae; small fish and amphibians.
General care
It has no particular demands as to water conditions and can withstand considerable fluctuations of temperature; for breeding, however, it needs an outside temperature of 18° – 20°C (65° – 68°F).
Notes
At spawning time the male builds a small nest consisting of a simple hole in the substrate, hollowed out with vigorous flicks of the tail. After a lively courtship, the female lays some thousands of eggs which are guarded by the male until they hatch.

PERCA FLUVIATILIS
Fam. Percidae

Common name
Perch.
Dimensions
Up to 45 cm (18 in).
Distribution
Inland and stagnant waters throughout Europe (apart from Spain), Asia Minor and northern Asia. A similar species occurs in North America.
Food
A voracious fish that feeds on insect larvae, small invertebrates and fish.
General care
It needs a large aquarium of 100 liters (22 gal) or more. There are no restrictions on water quality and it can be kept at temperatures of 15° – 25°C (60° – 77°F), although it will tolerate greater extremes – down to 5°C (41°F).
Notes
Aquarium breeding of this species is commonly practiced by placing two males and one mature female in a tank containing a few aquatic plants. The female will usually lay her eggs early in the morning, leaving a fertilized ribbon on the plants.

RHODEUS SERICEUS
Fam. Cyprinidae

Common name
Bitterling.
Dimensions
Up to 9 cm (3½ in).
Distribution
Europe (apart from Scandinavia); temperate Asia.
Food
In nature: omnivorous. In aquarium: dried fish food mixed with live food (worms, insect larvae, etc.).
General care
Adapts to unheated aquaria at temperatures of up to 22°C (72°F). Can be kept in small tanks filled with plants.
Notes
In nature the species lives in streams with muddy beds and plenty of vegetation. When spawning, the female lays her eggs inside the mantle cavity of large bivalve molluscs of the genus *Unio* (freshwater mussels). The fully developed young emerge from the exhalant syphon of the bivalve. A number of closely related species are available to aquarists.

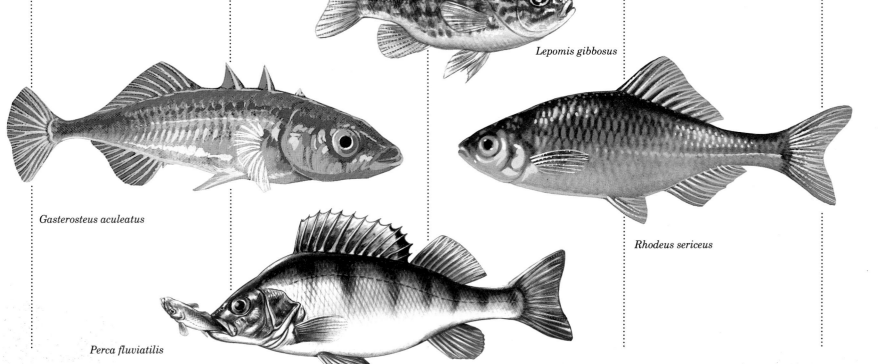

Lepomis gibbosus

Gasterosteus aculeatus

Rhodeus sericeus

Perca fluviatilis

TEMPERATE FRESH WATERS • 63

TROPICAL SEA WATERS

Tropical sea waters

Tropical sea waters

A bird's-eye view of a coral reef (bottom) provides only a foretaste of the surprises and delights in store for underwater divers (below).

Situated roughly between the tropics of Capricorn and Cancer, and extending for more than 160 million km² (62 million sq. miles), from the Red Sea to the Caribbean, is the earth's largest expanse of sea waters. One of the most fascinating and complex of all the world's ecosystems, it is distinguished principally by its corals and fish. The incredible variety of life to be found in what are biologically defined as tropical seas normally depends upon three parameters: temperature, light and suspended solids. Examination of the distribution of coral formations, in relation to the temperature of seas and oceans, reveals that madrepore, or hard corals – those organisms that actually make up coral reefs – becomes progressively less frequent, to the point of disappearance, as soon as the mean sea temperature fails to rise above 20°C (68°F) during the coldest period of the year. Moreover, it is easy to recognize how hard coral formations, in their characteristic variety of species, shapes and structures (atolls, barrier reefs, coral platforms) develop almost wholly along the shores of islands and continents and not in the depths of the oceans.

Light, of course, is closely related to temperature. Hard corals, in fact, need light in order to survive. The solar rays, therefore, do not only warm the water but also stimulate the growth in the corals of single-celled, symbiotic algae that play an all-important role in the development of the calcareous skeletons, the basis of coral formations. These algae, buried in the tissues of the polyps – the living parts of coral – can only carry out their photosynthetic activity if there is enough luminous energy. In extremely clear water, as is typical of many tropical seas, a sufficient amount of light penetrates to a depth of up to 50 m (165 ft) and in favourable instances even as far down as 90 m (300 ft). On the other hand, where the water is cloudier as a result of suspended sediment or plentiful concentrations of planktonic organisms, the light will be rapidly absorbed and coral reefs will only develop within 10 – 20 m (33 – 65 ft) of the surface.

This discussion of light reduction leads almost automatically to consideration of the third factor that controls life in the world of corals: suspended solids. Corals are sessile (i.e. immobile) animals. They are incapable of escaping by independent movement from variations in the conditions of their environment. If the layer of sediment is too thick, or the sedimentation rate too rapid, they will quickly become covered and be suffocated, slowing down metabolism, preventing photosynthesis of the symbiotic algae and causing their certain death. Water that is too cloudy and with too many solids in suspension will not readily be colonized by coral "spores" capable of building reefs of any notable size. This is clearly demonstrated along the northeastern Atlantic coasts of South America. Here the coral reef that stretches parallel to the shore from Rio de Janeiro to Florida is interrupted for 300 km (185 miles) opposite the mouth of the Amazon, because of the enormous quantity of sediment carried by the great river. This is what happens, too, in the case of marine sediments that lie too close to the surface and are continually held in suspension by waves and currents. From these brief observations it is possible to deduce the fundamental characteristics of a good tropical marine aquarium: a water temperature of 25° – 27°C (77° – 80°F); proper filtration; circulation that keeps the water clear but sufficiently stirred up and aerated; and lighting that will guarantee, as far as possible in the artificial aquarium environment, the survival of the polyps and the microalgae associated with them.

The appeal of the great coral reefs nevertheless resides chiefly in the fish that populate these waters. Displaying an incredible variety of colours and forms, they inhabit virtually every crack and cavity, swimming individually or in shoals in, around and above the coral formations. The limpid waters of the Great Barrier Reef alone accommodate more than 2,000 species, many of which are perfectly adapted to their exclusive habitat. This explains why a single coral formation of only a few square yards can contain a dozen or more different species, notwithstanding the competition for food and space that governs underwater life in tropical seas. In order to survive and derive the maximum benefit from the opportunities offered by their surroundings, coral fish have assumed forms and behaviour patterns that enable them to colonize every water layer down to considerable depths. The huge parrotfish have evolved mouths armed with strong teeth transformed into powerful beaks capable of crumbling the coral so as to extract food; and the elegant butterfly fish have soft snouts and small pointed

With careful selection and combination, the variety of life forms that are the hallmark of tropical seas can be recreated in the aquarium.

mouths, furnished with tiny teeth that can probe into the narrowest cracks in order to reach the minute invertebrates and polyps on which they feed.

Colours, too, which we tend to regard from a purely aesthetic viewpoint, assume a far more significant role when seen as adaptations to the environment. To us they may appear as no more than spectacular liveries, but in fact these are colours designed to confuse enemies, to make the wearers invisible or to issue precise and unmistakable messages: species, age, sex, mood (anger, fear, etc.) and state of health.

Anyone who comes into contact for the first time with the underwater life of a coral reef will find it virtually impossible to discern any rules and patterns or understand the complex relationships that exist among the innumerable species. Only constant observation and study, both in the wild and in the aquarium, will make sense of this ecosystem, so that eventually one comes to recognize the existence of territories and hierarchies among various species and individuals, and to discover the often unexpected links between different organisms, as is the case, for example, with anemones and clownfish, whose very existence depends on reciprocal aid. Then there are the cleaner fish who service other species that literally queue up to be rid of infesting parasites; the importance of such a function is such that if these tiny fish were to disappear from a barrier reef (as has been done experimentally), the entire fish population would be dramatically reduced in a matter of weeks.

These are perhaps familiar examples, but they do show that in order to build up an aquarium, in this case containing tropical marine species, it is not enough just to put the fish in a glass tank filled with water of suitable quality. The function of a successful aquarium, tropical or otherwise, has been perfectly defined by Konrad Lorenz in *King Solomon's Ring*, when he wrote: "... A man can sit for hours before an aquarium and stare into it as into the flames of an open fire or the rushing waters of a torrent. All conscious thought is happily lost in this state of apparent vacancy, and yet, in these hours of idleness, one learns essential truths about the macrocosm and the microcosm."

ABUDEFDUF TROSCHELLI
Fam. Pomacentridae

Common name
Panama sergeant major.
Dimensions
Up to 23 cm (9 in).
Distribution
From Gulf of California to Peru; Galapagos Islands.
Food
In nature: plankton and benthic invertebrates. In aquarium: a varied diet, including chopped fish and meat.
General care
It is a sturdy species but tends to become aggressive and bite its aquarium companions. Keep small specimens in an aquarium of 75 liters (16½ gal) or more, maintaining temperatures of 20° – 28°C (68° – 82°F).
Notes
Individuals of this species and its Atlantic relative *A. saxatilis* are robust fish, plentiful in shallow waters. During the spawning season the male becomes darker, almost bluish, and builds a nest on the cleaned portion of a rounded mass, where he fertilizes the eggs laid by the female. Other close relatives are *Nexillaris concolor* (= *A. declivifrons*) from the eastern Pacific, *A. abdominalis* from Hawaii and *A. bengalensis* from the Indo-Pacific region.

ACANTHURUS ACHILLES
Fam. Acanthuridae

Common name
Achilles tang, red-tailed surgeon.
Dimensions
Up to 25 cm (10 in).
Distribution
Tropical western and central Pacific, from Melanesia to Hawaii.
Food
In nature: principally herbivorous. In aquarium: chopped seafood (squid, bivalves, fish, etc.) mixed with vegetable matter and dried fish food containing vegetable matter.
General care
Will live happily together with other species. Young individuals are easiest to keep. Keep one fish in a tank of 75 liters (16½ gal) or more with various other species, at temperatures of 21° – 24°C (70° – 75°F).
Notes
This is a coastal species which normally lives close to coral reefs with strong-flowing channels.

ACANTHURUS LEUCOSTERNON
Fam. Acanthuridae

Common name
Powder blue surgeon.
Dimensions
Up to 22 cm (8½ in).
Distribution
Tropical Indo-Pacific region.
Food
As for *A. achilles*.
General care
Cannot live with individuals of the same species. Keep a young specimen with other species in an aquarium of 75 liters (16½ gal) or more, at temperatures of 22° – 28°C (72° – 82°F).
Notes
Aquarists consider this a highly coveted species. Avoid specimens from the Philippines that may have been captured with the aid of cyanide.

ACANTHURUS LINEATUS
Fam. Acanthuridae

Common name
Clown surgeon, blue-lined surgeon.
Dimensions
Up to 28 cm (11 in).
Distribution
Tropical Indo-Pacific region from East Africa to Tuamotu Islands.
Food
In nature: principally herbivorous; filamentous green algae. In aquarium: fish food containing plenty of vegetable matter mixed with chopped seafood, live *Artemia* and, if possible, green algae.
General care
Small individuals can be kept in a home aquarium although they tend to quarrel with one another. Keep a single specimen with other associated species and create a number of hiding places. The aquarium should hold at least 75 liters (16½ gal), at temperatures of 22° – 28°C (72° – 82°F).
Notes
This species is very abundant in shallow waters with rocky bottoms and off reefs down to a depth of 6 m (20 ft). It can be caught with traps and nets.

Acanthurus achilles

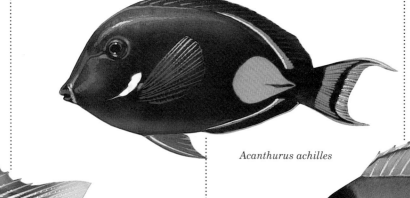

Abudefduf troschelli

Acanthurus leucosternon

Acanthurus lineatus

ACANTHURUS TRIOSTEGUS
Fam. Acanthuridae

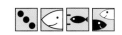

Common name
Striped surgeon.
Dimensions
Up to 15 cm (6 in).
Distribution
All tropical regions except western Atlantic.
Food
In nature: principally filamentous algae. In aquarium: chopped seafood mixed with vegetables such as broccoli, peas, spinach, etc; also dried fish food containing vegetable matter.
General care
Will live happily in an aquarium with one or more species. Keep several individuals together at temperatures of 22° – 28°C (72° – 82°F), in as large a tank as possible, preferably with a capacity of more than 400 liters (88 gal).
Notes
A. triostegus forms groups, sometimes quite large, when feeding on algae. It is abundant on the fringes of reefs at depths of up to 30 m (100 ft) and more. Young specimens are easily caught in pools. Adults may be taken in fixed nets or caught with hand nets late in the day when they gather for their last meal of the day. In the aquarium this species normally establishes a precise hierarchy when feeding, often resulting in the death of subordinate fish.

ADIONYX SUBORBITALIS
Fam. Holocentridae

Dimensions
Up to 25 cm (10 in).
Distribution
From Gulf of California and Mexico to Ecuador.
Food
In nature: carnivorous, principally small crustaceans. In aquarium: adapts well to a diet of chopped shrimp, squid and fish.
General care
A gregarious species, it chooses to hide in caves and crevices. Prefers temperatures of 20° – 27°C (68° – 80°F) but tolerates considerable fluctuations.
Notes
Individuals of this species hide in crevices during the day, normally in shallow water less than 3 m (10 ft) in depth, emerging at night, which is the only time they can be caught with nets by scuba divers. It is an aggressive species, not to be kept with others. The tank should be large (300 liters [66 gal]) and quite dimly lit so as not to disturb the fish with their dusk – nocturnal habits. The species, as a result of recent reclassification of the group, is described in some texts as *Sargocentron suborbitalis*.

AEOLISCUS STRIGATUS
Fam. Centriscidae

Common name
Razorfish, shrimpfish.
Dimensions
Up to 15 cm (6 in).
Distribution
Indo-Pacific region from Persian Gulf to Hawaii.
Food
In nature: carnivorous. In aquarium: live, newborn *Artemia*, small worms and zooplankton; dried fish food.
General care
Keep several specimens together in a tank without other species, with a sandy bed, natural or artificial plants, and long-spined sea urchins. The aquarium should contain 75 liters (16½ gal) and the sides and back covered to cause the least possible disturbance. Keep at temperatures of 22° – 28°C (72° – 82°F).
Notes
These are very attractive aquarium fish. Although delicate, if caught and handled carefully, they will adapt successfully to life in a tank. They usually swim vertically, head downward, and their typical defensive behaviour is to hide and camouflage themselves among the long spines of tropical sea urchins.

AMPHIPRION EPHIPPIUM
Fam. Pomacentridae

Common name
Fire clown, tomato clown, red saddleback clown.
Dimensions
Up to 10 cm (4 in).
Distribution
Andaman Sea, Sumatra, Java.
Food
In aquarium: feed on a varied diet comprising chopped shrimp, fish and bivalves mixed with live *Artemia* or algae, or flaked fish food containing vegetable matter.
General care
An ideal fish for beginners. The young can be kept in small 60-liter (13-gal) aquaria, but adults must be given a 200-liter (44-gal) tank. In captivity they do not need anemone hosts but the presence of these assists reproduction. Recommended temperatures are 28° – 29°C (82° – 84°F), although in the aquarium of the Scripps Institution (San Diego, California) they have been kept successfully at 23° – 25°C (74° – 77°F), as indicated by the fact that they spawn twice a month.
Notes
The young are notable for the presence of two vertical white bands, one behind the eye, the other in the middle of the body. Consequently this species is liable to be mistaken for the young of the related *A. frenatus*. In *A. ephippium*, however, the two bands disappear in time as the black spot on the body grows in size. After the formation of pairs, the adults turn aggressive.

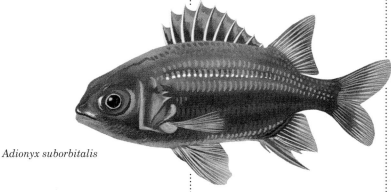

Adionyx suborbitalis

Acanthurus triostegus

Amphiprion ephippium

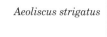

Aeoliscus strigatus

AMPHIPRION OCELLARIS
Fam. Pomacentridae

Common name
Common clown.
Dimensions
Up to 7.5 cm (3 in).
Distribution
Indo-Pacific region, from Ryukyu Islands to Australia, New Guinea and Philippines.
Food
Readily accepts *Artemia*, but gradually learns to live on dead food provided the size is appropriate.
General care
Young individuals require careful acclimation and only survive if large sea anemones are available.
Notes
Males remain smaller and slimmer than females. Unlike other species of *Amphiprion*, this clownfish is more markedly territorial, settling close to the point where the eggs have been laid, even after the removal of the anemone.

AMPHIPRION PERIDERAION
Fam. Pomacentridae

Common name
Salmon clown.
Dimensions
Up to 7.5 cm (3 in).
Distribution
Indo-Pacific region, including the South China Sea, Melanesia and Micronesia.
Food
Easily grows accustomed to frozen or freeze-dried food, though it is as well to alternate this with *Artemia*.
Notes
The species lives together with sea anemones of the genera *Stoichactis*, *Radianthus* and *Parasicyonis*, never straying far from them. The male is distinguished from the female by his smaller size and by having an orange band along the rear part of the dorsal fin and along the caudal fin. It is a delicate species, yet tranquil and easy to rear. The pairs are not aggressive towards their own young.

ANAMPSES MELEAGRIDES
Fam. Labridae

Common name
Yellow-tailed wrasse.
Dimensions
Up to 30 cm (12 in).
Distribution
Tropical Indo-Pacific region; it is not found in Hawaii and the eastern Pacific.
Food and general care
As for *Gomphosus varius*.
Notes
The male is violet in colour. The body scales display round white spots. The tail is yellow and violet with a dark edge. The species is fairly easy to rear. In nature it lives alone or in pairs, and this tendency persists in the aquarium. In small tanks it exhibits aggressiveness towards members of its own or related species, and towards other smaller fish. It needs hiding places and plenty of space for swimming.

ANISOTREMUS TAENIATUS
Fam. Haemulidae

Common name
Atlantic porkfish.
Dimensions
Up to 30 cm (12 in).
Distribution
From Cape San Luca (Mexico) to Panama.
Food
In nature: worms, brittle stars, crustaceans and molluscs; nocturnal. In aquarium: chopped shrimp, squid and bivalves. Feed at night.
General care
The young adapt well to living with other species. The adults need tanks of 1,000 liters (220 gal) or more and are usually unsuited to home aquaria. Maintain temperatures of 20° – 23°C (68° – 74°F).
Notes
The young differ from the adults in coloration. They are yellowish, with two black stripes on the flanks. Sometimes they behave as cleaner fish. A related species, *A. virginicus*, lives in the western Atlantic.

Amphiprion ocellaris

Amphiprion perideraion

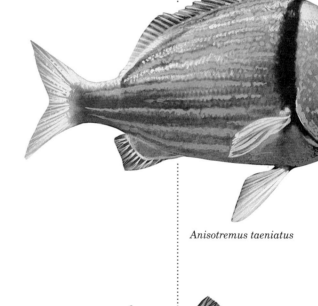

Anisotremus taeniatus

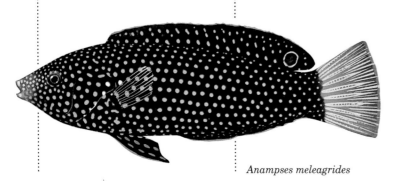

Anampses meleagrides

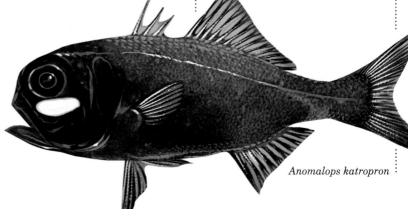

Anomalops katropron

ANOMALOPS KATROPRON
Fam. Anomalopidae

Common name
Lanternfish, flashlight fish.
Dimensions
Up to 30 cm (12 in).
Distribution
Not known precisely, but widespread in Indo-Pacific region.
Food
In nature: carnivorous, small fish and crustaceans. In aquarium: live *Artemia*, small live fish.
General care
Use an unilluminated aquarium in a dim corner of the room. Cover the back and sides with opaque board and use stones and tiles to build dark hiding places. A 10-cm (4-in) specimen should be kept in a tank of 200 liters (44 gal) or more, at a temperature of 20° – 24°C (68° – 75°F).
Notes
Some small specimens of *Anomalops*, caught in the Philippines, have been exported for aquarists. This species is less sturdy than *Photoblepharon* and its colour tends to be attenuated in captivity.

ANTENNARIUS HISPIDUS
Fam. Antennariidae

Common name
Frogfish.
Dimensions
Up to 20 cm (8 in).
Distribution
Tropical Indian Ocean and western Pacific.
Food
In nature: carnivorous. In aquarium: small live fish and any kind of fresh or frozen molluscs that it will take.
General care
Although it can be introduced to the aquarium with other species, it lives better with those of its own kind and when provided with live prey. Maintain water temperatures of 24° – 29°C (75° – 84°F).
Notes
The coloration of this species varies from yellowish to brown, with zebra-type stripes. Normally it inhabits calm zones of coral reefs where it attracts a wide range of prey.

ANTENNARIUS SANGUINEUS
Fam. Antennariidae

Common name
Bloody frogfish.
Dimensions
Up to 9 cm (3½ in).
Distribution
From Gulf of California to Peru.
Food
In nature: small fish and crustaceans. In aquarium: small live fish and adult *Artemia* mixed with chopped squid, shrimp and fish.
General care
Suitable for keeping with other species but take care to feed it regularly. Maintain water temperatures of 20° – 24°C (68° – 75°F) in an aquarium with rocky hiding places.
Notes
This species lives in dark crevices along rocky coasts. They are best seen with strong underwater lighting.

APOGON MACULATUS
Fam. Apogonidae

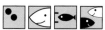

Common name
Flamefish.
Dimensions
Up to 13 cm (5¼ in).
Distribution
Tropical western Atlantic; widespread north to New England and south to Brazil.
Food
In nature: plankton and principally crustaceans. In aquarium: feeds voraciously, doing best on dried fish food mixed with chopped shrimp and live *Artemia*.
General care
Adapts easily to captivity. Several can be kept together in a tank with one or more other species. Create shelters and crevices. Maintain temperatures of 22° – 28°C (72° – 82°F).
Notes
This species differs from others to which it is closely related by having dark markings on the caudal peduncle and two white lines through the eye. It is widely distributed in shallow waters of the western Atlantic. Nocturnal by habit, it hides by day in caves and crevices, emerging at night to feed. It is found from the surface down to at least 120 m (400 ft). It spawns in early summer and mouth-broods its young.

Antennarius hispidus

Antennarius sanguineus

Apogon maculatus

TROPICAL SEA WATERS • 71

APOGON RETROSELLA
Fam. Apogonidae

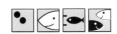

Common name
Banded cardinalfish.
Dimensions
Up to 10 cm (4 in).
Distribution
From Gulf of California to Oaxaca, Mexico.
Food
In nature: principally small crustaceans. In aquarium: live adult *Artemia*, chopped shrimp, squid and bivalves; also krill and dried fish food.
General care
Adapts to living with other tranquil species. Needs one or more separate cavities in which to hide. Keep at temperatures of 20° – 23°C (68° – 74°F).
Notes
Members of this species, like other cardinalfish, are nocturnal and hide by day in caves and crevices, usually sharing them with *Myripristis leiognathos*. They live on rocky reefs and along coasts to a depth of 60 m (200 ft). At night they emerge to feed on planktonic crustaceans; if there is no moon, they may venture into deeper water some distance from their refuge. Like other members of the family, *A. retrosella* is a mouth-brooder. Spawning has been observed in September.

APOLEMICHTHYS ARCUATUS
Fam. Pomacanthidae

Common name
Bandit angelfish.
Dimensions
Up to 18 cm (7 in).
Distribution
Hawaii.
Food
In nature: sponges. In aquarium: feed on a varied diet that includes chopped shrimp and squid, *Artemia* and flaked fish food containing vegetable matter.
General care
Keep in an average-sized aquarium of 300 liters (66 gal) or more at temperatures of 24° – 28°C (75° – 82°F); create caves and crevices as hiding places.
Notes
This is a species much coveted by aquarists who specialize in rare and unusual fish. It is essential to carry out various tests to determine the best conditions as regards space, lighting, diet and aquarium companions. In nature the species normally lives at depths of 12 – 50 m (40 – 165 ft) on stony bottoms. Individuals can easily be caught with hand nets.

AROTHRON MELEAGRIS
Fam. Tetraodontidae

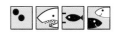

Common name
Golden puffer, spotted puffer.
Dimensions
Up to 30 cm (12 in).
Distribution
Tropical Pacific.
Food
In nature: carnivorous, feeding on a wide variety of invertebrates, from coral polyps to tunicates. In aquarium: chopped seafood.
General care
The species produces toxic mucus but can be kept successfully in a large aquarium, at a temperature of 22° – 28°C (72° – 82°F). Partial and regular water changes are recommended.
Notes
A. meleagris goes through a stage when it is bright yellow, the ecological significance of which is unknown. It is easy to catch but precautions must be taken because of the poisonous mucus it secretes when disturbed.

AROTHRON RETICULARIS
Fam. Tetraodontidae

Common name
Reticulated puffer.
Dimensions
Up to 42 cm (16½ in).
Distribution
India, Indo-Malayan peninsula, Philippines, New Guinea and northern Australia to Fiji.
Food
It is an omnivorous species, but in the aquarium it is essential to provide animals with shells (molluscs and crustaceans) so that the tooth plates can be kept functioning properly. The same applies to *A. meleagris*.
General care
It needs a sandy bottom with rocks and stones for hiding places, good lighting and a water temperature of 24° – 28°C (75° – 82°F).
Notes
Easily acclimated, this fish is a very entertaining species. It lives happily with others, but because of its territorial behaviour, it is inclined to be aggressive towards other Tetraodontidae. Like other specimens of the same genus, it will live for some time (5 – 7 years) in captivity.

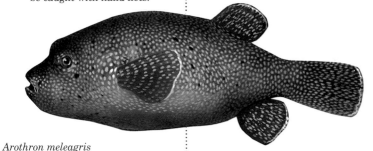
Arothron meleagris

Apogon retrosella

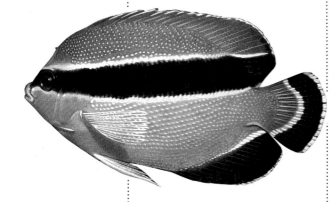
Apolemichthys arcuatus

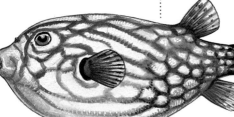
Arothron reticularis

ASPIDONTUS TAENIATUS
Fam. Blenniidae

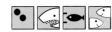

Common name
False cleaner fish, sabre-toothed blenny.
Dimensions
Up to 13 cm (5¼ in).
Distribution
Tropical Indo-Pacific region from Red Sea to Tuamotu Islands.
Food
In nature: sharks, fins and skin of bigger fish, worms and other invertebrates. In aquarium: small worms, chopped fish and other types of chopped seafood.
General care
It may share a tank with other species but needs to be fed well and promptly removed if it attacks other fish. Provide burrows or holes in which it can hide. The temperature should be 22° – 28°C (72° – 82°F).
Notes
This is a particularly interesting species for aquarists looking for a fish with an unusual behaviour pattern. It mimics perfectly the cleaner wrasse (*Labroides dimidiatus*), so that it can approach other fish without arousing suspicion; it then proceeds to rip off their scales or strips of skin rather than ridding them of parasites.

BALISTAPUS UNDULATUS
Fam. Balistidae

Common name
Undulate triggerfish.
Dimensions
Up to 30 cm (12 in).
Distribution
Tropical Indo-Pacific region, except Hawaii and North America.
Food
In nature: carnivorous, its diet including coral. In aquarium: chopped seafood such as shrimp, bivalves, squid and fish; provide fish food in pellet form.
General care
Individuals of this species sometimes have to be separated from their aquarium companions because of their aggressive nature. Try to put them in tanks with smaller species but be ready to move them if they become too combative. Keep temperatures at 22° – 28°C (72° – 82°F).
Notes
Commonly found on the outer slopes of coral reefs down to a depth of 40 m (130 ft).

BALISTES VETULA
Fam. Balistidae

Common name
Queen triggerfish.
Dimensions
Up to 50 cm (20 in).
Distribution
Tropical Atlantic and Indian Oceans.
Food
In nature: carnivorous, principally sea urchins. In aquarium: accepts any kind of fish food.
General care
Keep a single specimen in an aquarium containing other species. Create crevices for shelter at night. Recommended temperatures are 22° – 24°C (72° – 75°F).
Notes
This is one of the loveliest triggerfish, quiet with other species but aggressive towards those of its own kind. As adults they display sexual dimorphism; the males are brilliantly coloured and develop long appendages on the dorsal, anal and caudal fins. They have not been bred in captivity.

BALISTOIDES CONSPICILLUM
Fam. Balistidae

Common name
Clown trigger.
Dimensions
Up to 50 cm (20 in).
Distribution
Tropical Indo-Pacific region, from East Africa to Samoa.
Food
In nature: carnivorous; capable of feeding on hard-shelled invertebrates. In aquarium: chopped seafood, whole crabs and other types of fish food.
General care
Small specimens can live in an aquarium with other species but as they grow they may turn aggressive and have to be removed. Make crevices as shelters. Maintain temperatures of 22° – 28°C (72° – 82°F).
Notes
Lives on slopes and crags of coral reefs at depths of 3 – 40 m (10 – 135 ft). The significance of its markings is unknown, but because it turns and displays the marks on its abdomen when it is hunted, the theory is that they may serve to confuse predators.

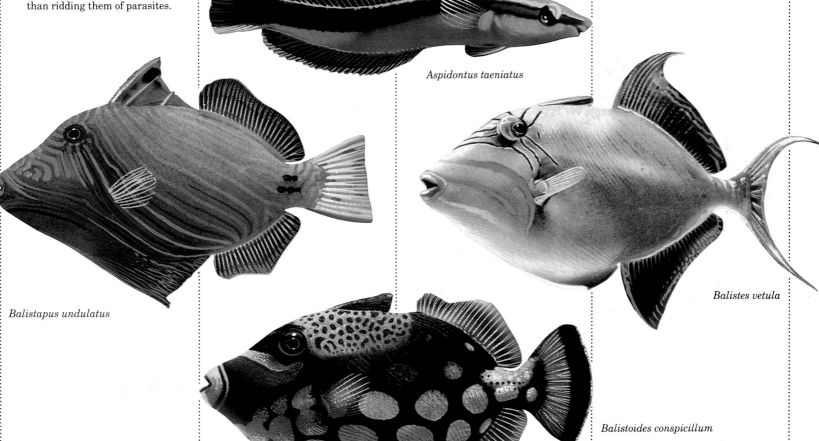

Aspidontus taeniatus

Balistapus undulatus

Balistes vetula

Balistoides conspicillum

TROPICAL SEA WATERS • 73

BOTHUS MANCUS
Fam. Bothidae

Common name
Pacific brill.
Dimensions
Up to 40 cm (16 in).
Distribution
Tropical Indo-Pacific region, from East Africa to Pitcairn Island and northward to Hawaii.
Food
In nature: carnivorous, feeding on crustaceans and benthic fish. In aquarium: chopped bivalves, squid and fish.
General care
Small specimens are suitable for keeping in domestic aquaria, even together with other species, in a tank provided with a zone of sand, at a temperature of 22° – 28°C (72° – 82°F).
Notes
The presence of these fish may often be overlooked in an aquarium containing a number of species. They are widely distributed in sandy areas of reefs and lagoons where the attentive observer may be able to catch them with a net.

CANTHIGASTER PUNCTATISSIMA
Fam. Tetraodontidae

Common name
Sharpnose puffer.
Dimensions
Up to 9 cm (3½ in).
Distribution
From Gulf of California to Panama.
Food
In nature: not known. In aquarium: does well on a diet of chopped seafood with an addition of live *Artemia*.
General care
Will adapt to life with other varieties but may be boisterous with members of its own species. Needs a large aquarium of 400 liters (88 gal) or more. Maintain a temperature of 20° – 24°C (68° – 75°F). Partial and regular water changes are recommended.
Notes
This species is very common in the Gulf of California, along the southern shores of the peninsula of Lower California. It is as slow swimmer, normally remaining close to its lair, and may be caught easily with a net. This species, too, secretes toxic mucus when touched or disturbed. Many similar species are commercially available, including *C. valentini* from the eastern Indo-Pacific.

CENTROPYGE ARGI
Fam. Pomacanthidae

Common name
Cherubfish, pygmy angelfish.
Dimensions
Up to 6.5 cm (2¾ in).
Distribution
Bermuda, West Indies and southern Gulf of Mexico.
Food
In nature: principally algae. In aquarium: include flaked fish food containing greens and vegetable matter.
General care
A good aquarium fish, it can be kept in a tank of 100 liters (22 gal) or more, at a temperature of 20° – 24°C (68° – 75°F).
Notes
This attractive species is relatively rare in shallow waters and often favours stony areas at depths of 30 – 70 m (100 – 220 ft). The yellow head coloration of young and juvenile specimens disappears in adults. As happens in the wild, individuals of *C. argi* tend to demarcate territory that is then defended against others of the same species. They need cracks and crevices for refuge.

CENTROPYGE BISPINOSUS
Fam. Pomacanthidae

Common name
Coral beauty, purple angel.
Dimensions
Up to 8 cm (3 in).
Distribution
Indo-Pacific region from East Africa eastward to Gilbert Islands, north to Japan and south to Lord Howe Island.
Food
In nature: algae. In aquarium: as for *C. flavissimus*.
General care
Keep in an aquarium of 200 liters (44 gal) or more, with plenty of hiding places, maintained at a temperature of 24° – 28°C (75° – 82°F).
Notes
Lives close to lair on coral or stony beds, at depths of 5 – 50 m (16 – 165 ft). As a rule, available specimens come from the Philippines, where they may be caught with chemical substances. According to De Graaf, these individuals have a slightly different colour and are offered under the name of *C. kennedy*, which is not scientifically recognized. Small fish acclimate more easily if kept in aquaria with filamentous algae.

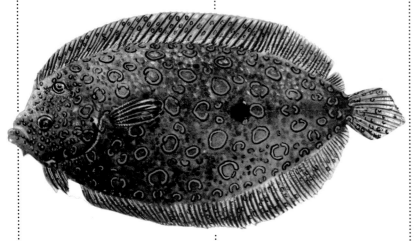

Bothus mancus

Centropyge argi

Centropyge bispinosus

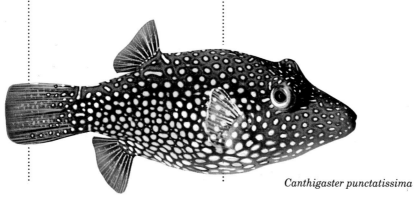

Canthigaster punctatissima

CENTROPYGE FLAVISSIMUS
Fam. Pomacanthidae

Common name
Lemonpeel angelfish.
Dimensions
Up to 10 cm (4 in).
Distribution
Indo-Pacific region from Great Barrier Reef to Hawaii and the Marquesas.
Food
In nature: algae. In aquarium: algae and other vegetable matter (greens, flaked fish food) mixed with chopped shrimp and squid, and live *Artemia*.
General care
This species, although it feeds mainly on algae in the wild, must be given a varied diet in the aquarium. It needs a tank of at least 200 liters (44 gal), at a temperature of 24° – 28°C (75° – 82°F), with plenty of hiding places.
Notes
This is the most delicate of the *Centropyge* species described. Young individuals are distinguished by a dark blue-bordered spot in the center of the flanks beneath the lateral line. The species is common in the central Pacific, in shallow water to a depth of 3 – 20 m (10 – 65 ft). A close relative, *C. heraldi*, lacks the blue ring around the eye.

CENTROPYGE LORICULUS
Fam. Pomacanthidae

Common name
Flame angelfish.
Dimensions
Up to 12 cm (5 in).
Distribution
Principally tropical central Pacific, including Johnston Island, but also found in New Guinea.
Food
In nature: omnivorous. In aquarium: normal varied diet.
General care
A robust species, excellent for the aquarium. Young individuals can be kept in a 75-liter (16½-gal) aquarium, but a larger one is even better. Maintain temperatures of 24° – 28°C (75° – 82°F).
Notes
Lives on the outer edge of barrier reefs, often at some depth, close to beds characterized by large, solitary coral formations. Territorial, like all the *Centropyge* species, it does well in a large tank or one with plenty of hiding places where it can remain for long periods undisturbed. Sometimes the species is described as *C. flammeus*, but this is a synonym that is not scientifically accepted.

CENTROPYGE POTTERI
Fam. Pomacanthidae

Common name
Russet angelfish.
Dimensions
Up to 10 cm (4 in).
Distribution
Hawaii.
Food
In nature: algae and invertebrates. In aquarium: accepts various types of live food, fresh, frozen or dried; mix with vegetable matter.
General care
Normally it adapts quite well to captivity but it needs plenty of hiding places. Use an aquarium of 200 liters (44 gal) or more, at temperatures of 20° – 25°C (68° – 77°F).
Notes
Sometimes this species turns aggressive in the aquarium, so remove it if it creates problems. It is one of the most difficult *Centropyge* species to keep successfully, mainly because it requires live food and an abundance of algae. Only after long acclimation will it accept frozen or freeze-dried food. Many specimens, after introduction to the tank, may not feed for a long time.

CEPHALOPHOLIS ARGUS
Fam. Serranidae

Common name
Argus grouper.
Dimensions
Up to 40 cm (16 in).
Distribution
Tropical Indo-Pacific region from South Africa to Tuamotu archipelago and Hawaii (introduced).
Food
In nature: carnivorous, feeding on small fish and crustaceans both by day and night. In aquarium: fish, squid and crustaceans.
General care
Easy to keep and will live happily with other large fish. Use the largest possible aquarium with hiding places among the rocks and coral. Maintain water temperatures at 20° – 30°C (68° – 86°F).
Notes
This species is most widely distributed in zones with plenty of coral. Adults are found from the surface down to a depth of 40 m (135 ft) or more. The young usually live in protected, shallower areas. They grow rather slowly in captivity but survive for many years if the aquarium offers ideal conditions. The species was introduced to Hawaii in 1956 from the Society Islands.

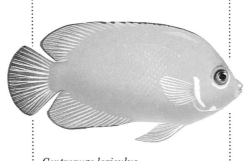

Centropyge loriculus

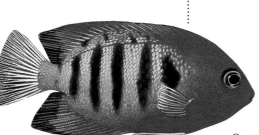

Centropyge flavissimus

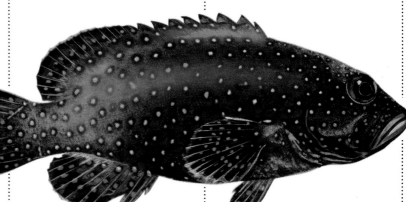

Centropyge potteri

Cephalopholis argus

TROPICAL SEA WATERS

CEPHALOPHOLIS MINIATUS
Fam. Serranidae

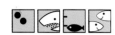

Common name
Coral trout, red grouper.
Dimensions
Up to 45 cm (18 in).
Distribution
Tropical Indo-Pacific region from Red Sea to Tahiti; absent in Hawaii.
Food
In nature: carnivorous. In aquarium: fish, shrimp and crustaceans.
General care
Lives happily in a tank with other species, although they should not be too small as they may be eaten. Prepare shelters among the rocks and, if desired, the coral. Maintain temperatures of 21° – 28°C (70° – 82°F).
Notes
This is a vividly coloured fish which adds a lively note to an aquarium containing several species. As happens among many tropical groupers, *C. miniatus* changes livery, becoming progressively darker red with more blue smaller spots. A territorial fish, it tends to settle in a separate zone of the tank along with fish of a smaller size. Although widely found in coral reefs, it is seldom caught with hook and line. Some species look much the same, notably *Plectropoma maculatum* and *Variola louti*, but these do not have a rounded tail.

CHAENOPSIS ALEPIDOTA
Fam. Chaenopsidae

Common name
Orange blenny pike.
Dimensions
Up to 15 cm (6 in).
Distribution
Gulf of California and Pacific coasts from Southern California to Cape San Luca (Mexico).
Food
In nature: carnivorous; feeds mainly on small crustaceans but also small worms and fish. In aquarium: live *Artemia*, worms, chopped seafood, dried fish food.
General care
Adapts to living in a tank with one or more small, peaceful species. Insert rocks fitted with polychaete holes or use artificial tubes (e.g. cane) as refuges. A number of these fish may be kept in a 200-liter (44-gal) tank, but the males are territorial.
Notes
As a rule, it lives in polychaete holes close to reefs, on a sandy or pebbly bed, from the surface down to a depth of 20 m (65 ft). Scuba divers can catch them simply by collecting rocks and shells with polychaete holes containing the fish or by digging up the holes that accommodate them.

CHAETODIPTERUS ZONATUS
Fam. Ephippididae

Common name
Spadefish.
Dimensions
Up to 65 cm (26 in).
Distribution
From San Diego (California) to Peru.
Food
In nature: omnivorous. In aquarium: pieces of fish, squid, bivalves or shrimp.
General care
They adapt well to living in an aquarium with other species, but the adults need a very large tank of 2,000 liters (440 gal) or more. Prepare a sandy bed and, if desired, scatter some stones around. Maintain a temperature of 20° – 23°C (68° – 74°F).
Notes
These are shy, placid aquarium fish. Aquarists prefer young individuals. As a rule, they live in groups in bays with sandy or stony bottoms, occupying shallow water to a depth of about 45 m (150 ft). At night they often come in to shore (where they can be fished with trawl nets) or collect around posts and jetties to feed on organisms attached to them (in which case they may be caught by scuba divers with torches or with hand nets).
They reproduce by laying eggs, apparently in pairs, just beneath the water surface. One closely related species, *C. faber*, lives in the Atlantic.

CHAETODON ACULEATUS
Fam. Chaetodontidae

Common name
Poey's butterfly fish.
Dimensions
Up to 10 cm (4 in).
Distribution
Gulf of Mexico, Florida and northern Caribbean.
Food
In nature: feeds on live invertebrates in coral and rock cracks. In aquarium: accepts a wide variety of food.
General care
It is a robust species which can be kept in an aquarium of 200 liters (44 gal) or more, at temperatures of 20° – 27°C (68° – 80°F).
Notes
These fish normally live in fairly deep water at 20 – 50 m (65 – 165 ft), often near rocky reefs. Adult individuals live alone or in pairs. It is the most widely distributed butterfly fish of the Caribbean, very delicate and difficult to raise. The long-snout shape reflects its food preferences, chiefly invertebrates which, nevertheless, should not be kept in the same tank. It is sometimes described by the name *Prognathodes aculeatus*.

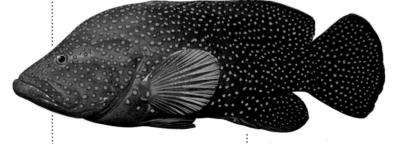

Cephalopholis miniatus

Chaetodon aculeatus

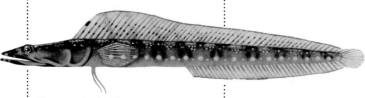

Chaenopsis alepidota

Chaetodipterus zonatus

CHAETODON ARGENTATUS
Fam. Chaetodontidae

Common name
Silver butterfly fish.
Dimensions
Up to 20 cm (8 in).
Distribution
From southern Japan to Philippines.
Food
In nature: omnivorous. In aquarium: accepts a wide range of food. Give live food occasionally.
General care
Adapts well to captivity. Use an aquarium of at least 200 liters (44 gal). It lives best at temperatures of 20° – 28°C (68° – 82°F).
Notes
This omnivorous species lives at depths of 5 – 20 m (16 – 65 ft). It is found around cliffs in the colder areas of its range, and on barrier reefs in warmer regions. It is a relatively placid species. The tank must contain plenty of crevices and be well illuminated, possibly with some sunlight.

CHAETODON AURIGA
Fam. Chaetodontidae

Common name
Threadfin butterfly fish.
Dimensions
Up to 23 cm (9 in).
Distribution
Indo-Pacific region from Hawaii to Red Sea.
Food
In nature: omnivorous, including coral, crustaceans and worms. In aquarium: accepts a wide range of fresh, frozen or dried food.
General care
Small individuals adapt to a tank of 150 liters (33 gal) or more, at a temperature of 23° – 28°C (74° – 82°F).
Notes
This is a popular aquarium fish which will live for some years in good conditions and with a proper diet. In adult specimens there is a long, yellow thread-like ray on the dorsal fin, above its characteristic dark eye spot. Some experts recognize the existence of two subspecies: *C. auriga auriga* from the Red Sea, whose adults lack the eye spot, and *C. auriger setifer*, found in the Indian and Pacific Oceans, in whom the dorsal eye spot persists throughout life.

CHAETODON COLLARE
Fam. Chaetodontidae

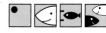

Common name
Pakistani butterfly fish.
Dimensions
Up to 16 cm (6¼ in).
Distribution
Indo-Pacific region from Persian Gulf to Philippines; absent in Australia.
Food
In nature: coral polyps and live invertebrates on coral. In aquarium: provide a varied diet, including crustaceans and, if possible, fish food in pellet form.
General care
Use a large aquarium of 400 liters (88 gal) or more and maintain a temperature of 24° – 28°C (75° – 82°F).
Notes
This species lives both on coral reefs and rocks, at depths of 4 – 15 m (13 – 50 ft). Individuals caught in rocky zones adapt better to captivity than those fished off reefs where food consists exclusively of coral polyps. Make sure that specimens from the Philippines have not been caught with chemical products such as cyanide.

CHAETODON EPHIPPIUM
Fam. Chaetodontidae

Common name
Saddleback butterfly fish.
Dimensions
Up to 12 cm (5 in).
Distribution
Pacific Ocean from southern Japan to Australia and Thailand, and eastward to Hawaii.
Food
In nature: omnivorous, including coral polyps. In aquarium: provide a wide variety of food.
General care
Recommended to more experienced aquarists. Use an average-sized tank of 300 liters (66 gal) or more, maintaining temperatures at 24° – 28°C (75° – 82°F).
Notes
Very young individuals lack the long terminal filament on the dorsal fin that typifies the adults, but have a dark band over the eyes which becomes less conspicuous with age.

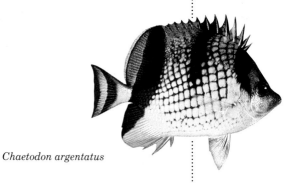

Chaetodon argentatus

Chaetodon auriga

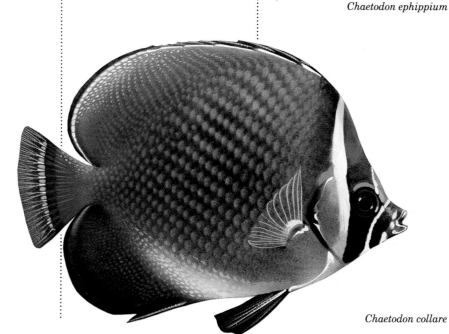

Chaetodon ephippium

Chaetodon collare

CHAETODON FALCIFER
Fam. Chaetodontidae

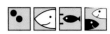

Common name
Falcate butterfly fish.
Dimensions
Mature adults 13 – 18 cm (5¼ – 7 in).
Distribution
Southern California (rare) and Galapagos Islands.
Food
In nature: feeds on organisms floating on the bottom, in rocky zones, including cliff walls. In aquarium: lives happily on a mixed diet of fish and molluscs with the occasional addition of *Artemia*; also takes dried fish food.
General care
It is an extremely hardy aquarium species, living best at temperatures of 13° – 21°C (56° – 70°F). Some individuals in the aquarium of the Scripps Institution of San Diego, California, continue to prosper after many years in captivity.
Notes
This is a cold-water species; in tropical regions it lives below 35 m (100 ft) where the temperature is low, but it may come to the surface more often in cool, temperate zones. Large shoals have been observed deeper than 75 m (200 ft) off the tip of Lower California, the center of its habitat, but it is easier to find them off the islands of San Benito and Guadelupe.

CHAETODON HUMERALIS
Fam. Chaetodontidae

Common name
Banded butterfly fish.
Dimensions
Adults 10 – 15 cm (4 – 6 in), maximum 25 cm (10 in).
Distribution
From Gulf of California to Peru and Galapagos Islands; rarely north of California.
Food
In nature: carnivorous. In aquarium: chopped squid and shrimp, flaked aquarium food with chopped fish and *Artemia* to vary the diet.
General care
Tolerates a wide temperature range, but 20° – 24°C (68° – 75°F) is recommended.
Notes
This is the most common species in the eastern Pacific. It lives in pairs or small groups, in shallow water, at depths of 1 – 12 m (3 – 40 ft), close to cliffs and wharfs, sometimes venturing near beaches. Adults and young have the same coloration. A scuba diver with a torch can easily catch them at night; by day they often get caught up in the drag nets of vessels fishing for prawns.

CHAETODON KLEINI
Fam. Chaetodontidae

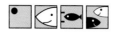

Common name
Klein's butterfly fish.
Dimensions
Up to 13 cm (5¼ in).
Distribution
Indo-Pacific region from South Africa to Hawaii, including Red Sea.
Food
In nature: omnivorous. In aquarium: accepts a wide variety of fish, live or frozen, and invertebrates; flaked fish food.
General care
Keep in an aquarium of 200 liters (44 gal) or more at a temperature of 24° – 28°C (75° – 82°F).
Notes
Adults are normally found at depths of 25 – 55 m (85 – 180 ft) in Hawaii, and at lesser depths in the Philippines. Young individuals live in shallower water in both regions. Once adapted to captivity, this species may be very long-lived. Even though it is not highly coloured, the small size, active nature and sturdiness make it a favourite aquarium choice.

CHAETODON LUNULA
Fam. Chaetodontidae

Common name
Racoon butterfly fish.
Dimensions
Up to 21 cm (8¼ in).
Distribution
Tropical Indo-Pacific region from East Africa to Hawaii and Marquesas. Absent in Red Sea.
Food
In nature: omnivorous. In aquarium: provide a varied diet including ordinary fish foods.
General care
This is a sturdy species (for example, in the Steinhart Aquarium of San Francisco it usually lives for 7 or more years). Use a tank of 200 liters (44 gal) or more at a temperature of 22° – 28°C (72° – 82°F).
Notes
It normally lives in the shallow waters of reefs down to 5 m (16 ft). The young often appear in pools and are particularly suited to aquarium life. Given its preference for anemones and coral polyps, it is best not to house them along with invertebrates. The dark yellow-edged spot on the terminal part of the dorsal fin in young specimens tends to be transformed gradually into a vertical band. It is timid in the wild and lives alone, in pairs or groups. It is easily caught at night with nets or traps.

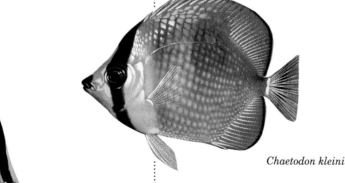

Chaetodon kleini

Chaetodon falcifer

Chaetodon humeralis

Chaetodon lunula

CHAETODON OCELLATUS
Fam. Chaetodontidae

Common name
Eyed butterfly fish.
Dimensions
Up to 16 cm (6¼ in).
Distribution
Tropical western Atlantic from Florida to Brazil. Seasonally it moves northward as far as Massachusetts and Nova Scotia.
Food
In nature: omnivorous. In aquarium: accepts a large variety of fresh, frozen and dried food or live organisms.
General care
It is a robust species. Although several small specimens can be kept in an 80-liter (17½-gal) aquarium, a tank of at least 200 liters (44 gal) is preferable. Maintain temperatures of 20° – 26°C (68° – 79°F).
Notes
This freely available species is much collected by American aquarists. During summer and early autumn it is commonly found in the northern part of its range, carried there in the form of eggs and larvae by the Gulf Stream. The young are characterized by a black stripe at the rear, which in the adults is reduced to a spot on the dorsal fin.

CHAETODON ORNATISSIMUS
Fam. Chaetodontidae

Common name
Ornate coralfish.
Dimensions
Up to 18 cm (7 in).
Distribution
Indo-Pacific region from Hawaii to Sri Lanka.
Food
In nature: coral polyps. In aquarium: *Artemia*, fish food in flake and pellet form, chopped shrimp and squid.
General care
Keep in a large aquarium of 400 liters (88 gal) or more at temperatures of 24° – 28°C (75° – 82°F). Recommended only to expert aquarists.
Notes
This marvellous species, though timid, lives in coral reefs to a depth of up to 15 m (50 ft). It is difficult to keep because it feeds on coral polyps. Feed frequently, twice or three times daily, and always have fish food in pellets handy. The aquarist who raises this species successfully can be justifiably proud.

CHAETODON PAUCIFASCIATUS
Fam. Chaetodontidae

Common name
Braided butterfly fish.
Dimensions
Up to 14 cm (5½ in).
Distribution
Red Sea and Gulf of Aden.
Food
In nature: omnivorous. In aquarium: normal varied diet.
General care
Install coral or stony hiding places in an aquarium of 250 liters (55 gal) or more, maintaining temperatures of 24° – 28°C (75° – 82°F).
Notes
This species may be hard to obtain. In nature it lives on coral reefs and in stony areas down to a depth of 30 m (100 ft). A close relative, *C. xanthurus*, is often exported from the Philippines, while two other similar species, *C. mertensii* (with orange spots) and *C. madagascariensis*, live in the Indo-Pacific region. All of them are beautiful and sturdy aquarium species.

CHAETODON RAFFLESI
Fam. Chaetodontidae

Common name
Raffles butterfly fish.
Dimensions
Up to 15 cm (6 in).
Distribution
Indo-Pacific region from Sri Lanka to Society Islands.
Food
In nature: coral polyps and small invertebrates. In aquarium: does well without coral polyps; provide a varied diet of *Artemia*, chopped shrimp and squid; flaked, dried and frozen fish food.
General care
Keep in an aquarium of 250 liters (55 gal) or more at temperatures of 24° – 28°C (75° – 82°F).
Notes
This species lives in fairly shallow water, normally in association with coral, at depths of 1 – 10 m (3 – 33 ft). As a rule it comes from the Philippines, so before buying, make sure it has not been caught with chemical products.

Chaetodon paucifasciatus

Chaetodon ocellatus

Chaetodon ornatissimus

Chaetodon rafflesi

CHAETODON TINKERI
Fam. Chaetodontidae

Common name
Tinker's butterfly fish.
Dimensions
Up to 15 cm (6 in).
Distribution
Hawaii.
Food
In nature: various benthic and planktonic animals. In aquarium: usually accepts a wide variety of food such as fish, shrimp, other invertebrates and flaked fish food.
General care
Keep in an aquarium of 300 liters (66 liters) or more at temperatures of 20° – 24°C (68° – 75°F).
Notes
This species, which lives in deep water, is neither observed nor caught frequently, but, like *C. falcifer*, it does accustom itself easily to captivity. It lives at depths of 40 – 75 m (135 – 230 ft) and can be kept at room temperature. When caught, it must be carefully degassed by means of a hypodermic needle and brought to the surface in a plastic bag filled with sea water from the same depth so as to avoid any harmful temperature fluctuations.

CHAETODON ULIETENSIS
Fam. Chaetodontidae

Common name
Saddled butterfly fish.
Dimensions
Up to 15 cm (6 in).
Distribution
Australia, New Guinea and western Pacific, including Tahiti and Fiji.
Food
In nature: omnivorous. In aquarium: accepts a wide variety of food; provide live food occasionally.
General care
As for *C. tinkeri*.
Notes
This species is frequently mistaken for its close relative, *C. falcula*, found only in the Indian Ocean, differing from it by its smaller area of yellow to the rear part of the body. Although rarely available to aquarists, it is an excellent fish for the home aquarium. As a rule it lives in association with coral at depths of 5 – 10 m (16 – 33 ft).

CHAETODONTOPLUS MESO-LEUCOS   Fam. Pomacanthidae

Common name
Variegated angelfish.
Dimensions
Up to 17 cm (6½ in).
Distribution
Western Pacific and Indian Ocean, from southern Japan to Indonesia.
Food
In nature: principally sponges and tunicates. In aquarium: give a varied diet containing chopped sea food, live *Artemia*, vegetable matter and flaked fish food.
General care
Recommended for expert aquarists, only because its food needs are not fully known. Use an aquarium of 300 liters (66 gal) or more, maintaining temperatures of 24° – 28°C (75° – 82°F).
Notes
This member of the Pomacanthidae, although very similar to those of the Chaetodontidae, possesses a well-developed opercular spine (a character of marine angelfish). It is found mainly in the Philippines, Indonesia and New Guinea, normally living along barrier reefs at depths of 2 – 20 m (7 – 65 ft). It is a solitary species, generally easy to approach and catch.

CHEILODIPTERUS MACRODON
Fam. Apogonidae

Common name
Eight-banded cardinalfish.
Dimensions
Up to 20 cm (8 in).
Distribution
Tropical Indo-Pacific region from East Indies to Society Islands; absent in Hawaii.
Food
In nature: small fish, large planktonic animals and soft-bodied benthic organisms. In aquarium: chopped fish, shrimp and squid mixed with small fish such as guppies.
General care
Needs a large tank of 200 liters (44 gal) or more with many hiding places among the rocks and coral. Can live with other species. Temperatures should be 22° – 28°C (72° – 82°F).
Notes
This genus, and particularly this species, is notable for possessing prominent canine teeth. Because *C. macrodon* is one of the larger Apogonidae, it is not as common as other species. It is found on coral reefs and in lagoons and may easily be caught at night with hand nets.

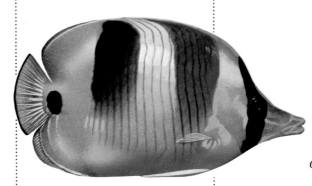

Chaetodon ulietensis

Chaetodon tinkeri

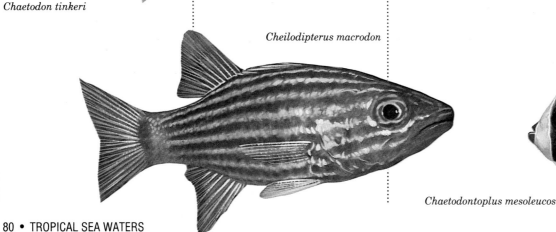

Cheilodipterus macrodon

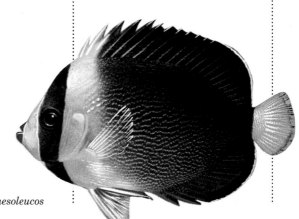

Chaetodontoplus mesoleucos

CHELMON ROSTRATUS
Fam. Chaetodontidae

Common name
Copper band butterfly fish.
Dimensions
Up to 20 cm (8 in).
Distribution
Indian Ocean and western Pacific, including Australia, Philippines, India and South Africa.
Food
In nature: feeds on living organisms in rock crevices, such as crustaceans, worms and other small invertebrates. In aquarium: chopped shrimp, chopped squid tentacles, krill, worms, flaked fish food, *Artemia*; initially it may need live food.
General care
Easily obtainable. Normally it is aggressive towards members of its own kind but it adapts to living with other species. Temperatures of 24° – 28°C (75° – 82°F) are recommended.
Notes
The fish has the long snout typical of the Chaetodontidae which feed on live invertebrates, inhabiting cracks and crevices of coral. It lives on coral reefs at depths of 1 – 10 m (3 – 33 ft). It prefers live food such as *Artemia*, worms and plankton, but may become accustomed to dead food. The species generally comes from the Philippines where it is sometimes caught with cyanide.

CHROMIS ATRILOBATA
Fam. Pomacentridae

Common name
Scissors chromis.
Dimensions
Up to 7.5 cm (3 in).
Distribution
From Gulf of California to Peru and Galapagos Islands.
Food
In nature: zooplankton. In aquarium: chopped fish, bivalves and squid.
General care
This is a good aquarium species. Since it is naturally gregarious, it does best with others of its own kind. Create shelters in a medium or large tank, at temperatures of 20° – 24°C (68° – 75°F).
Notes
This species lives in large groups over rocky beds and feeds on zooplankton. When not feeding, and during the night, it withdraws into caves and rock crevices. It is most commonly found at depths of 3 – 25 m (10 – 80 ft). A peaceful fish, it lives happily with others of its kind or together with other species. This is characteristic of almost all the *Chromis* species.

CHROMIS CAERULEA
Fam. Pomacentridae

Common name
Green chromis.
Dimensions
Up to 10 cm (4 in).
Distribution
Indo-Pacific region from East Africa to Tuamotu Islands.
Food
In nature: zooplankton. In aquarium: a wide variety of food, including chopped fish, shrimp and squid, and *Artemia*.
General care
It is a hardy, gregarious species, active and normally unaggressive in the aquarium. Keep at temperatures of 22° – 28°C (72° – 82°F).
Notes
Extremely abundant in coral atolls where it congregates in dense shoals, taking refuge among the branches of coral when danger threatens. The young in particular stick quite closely to the reefs. When spawning, the species becomes territorial, so that if the aquarium is not large enough, the males may turn aggressive, causing the deaths of some of the weaker contenders. *C. caerulea* should be kept together with smaller fish or invertebrates.

CHROMIS CYANEA
Fam. Pomacentridae

Common name
Blue chromis.
Dimensions
Up to 13 cm (5¼ in).
Distribution
Bermuda, Florida and West Indies.
Food
In nature: zooplankton. In aquarium: chopped foods.
General care
A gregarious and unaggressive species.
Notes
This peaceful fish forms dense shoals among choral reefs, living at depths of 10 – 25 m (33 – 80 ft). It is not a timid species and will seek shelter among the coral only if alarmed. Their tank should be provided with a strong water aeration system, so as to simulate the conditions of its natural environment. When spawning, the male becomes territorial, choosing a zone to attract and court the female. First he changes colour, turning black on the back and white on the belly, then he moves swiftly towards his mate, swimming around her before returning to his territory. The male's movements become increasingly frantic until the female eventually follows him and lays her eggs in the selected place.

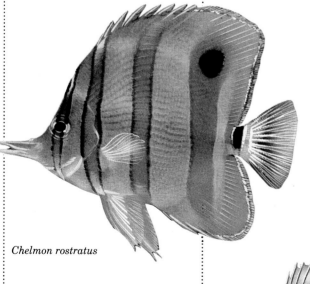

Chelmon rostratus

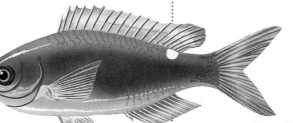

Chromis atrilobata

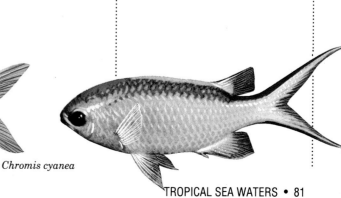

Chromis caerulea

Chromis cyanea

CHROMIS LIMBAUGHI
Fam. Pomacentridae

Common name
Limbaugh's chromis.
Dimensions
Up to 9 cm (3½ in).
Distribution
Central and southern Gulf of California.
Food
In nature: principally plankton living on the seabed. In aquarium: adapts to various diets. Try chopped fish, shrimp and squid, mixed with live and frozen *Artemia* and fish food.
General care
Normally tranquil, it lives happily in medium and large aquaria at temperatures of 20° – 23°C (68° – 74°F).
Notes
In nature they are highly colourful but in the aquarium they tend to lose their colours rapidly. As happens with many members of the genus *Chromis*, the young have a different coloration to the adults. Those of *C. limbaughi*, in fact, have yellow tints on the upper part of the body which disappear in the adults, who are violet – blue. It is a benthic species, usually found in rocky zones more than 15 m (50 ft) down. It is abundant among the islands around La Paz (Californian peninsula).

CHROMIS SCOTTII
Fam. Pomacentridae

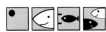

Common name
Purple chromis.
Dimensions
Up to 10 cm (4 in).
Distribution
Tropical western Atlantic.
Food
In nature: plankton living near the seabed. In aquarium: adapts to various foods. Does well on chopped shrimp and squid, and occasionally live *Artemia*.
General care
A good aquarium fish. The young are more brilliantly coloured, tending to gather in shoals, and are less quarrelsome than the adults. Small individuals can be kept in 150-liter (33-gal) tanks, at temperatures of 21° – 26°C (70° – 79°F).
Notes
This species normally lives near the bottom, in small groups at depths of 25 – 40 m (80 – 135 ft), although the young may venture up to 6 m (20 ft) in summer. Young fish are bright blue or purple. The adults have a blue half-moon mark above the eye.

CHRYSIPTERA CYANEA
Fam. Pomacentridae

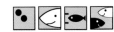

Common name
Blue devil.
Dimensions
Up to 8 cm (3 in).
Distribution
Indo-Australian archipelago to Samoa.
Food
In nature: omnivorous, feeding on algae, plankton and, above all, crustaceans and tunicates. In aquarium: live *Artemia* and chopped seafood, mixed with dried fish food containing vegetable matter.
General care
This is one of the less aggressive of the Pomacentridae. It can be kept with other species but is also attractive in aquaria reserved for them alone. Prepare plenty of refuges among the coral. Keep several specimens together in an aquarium of 100 liters (22 gal) or more, at temperatures of 22° – 28°C (72° – 82°F).
Notes
This is one of the best aquarium species of the Pomacentridae. It is colourful, active and markedly less quarrelsome than others of its family. Mature males are bicoloured with a yellow abdomen and tail. Females lack the yellow tail but have a black spot at the base of the dorsal fin. Nevertheless, coloration and pattern vary considerably according to the area of distribution; males from Palau have vivid orange markings rather than yellow. The species is found plentifully in rocky zones. *C. cyanea* can be caught with a fixed net or fish trap.

CIRRHITICHTHYS OXYCEPHALUS
Fam. Cirrhitidae

Common name
Coral hawkfish.
Dimensions
Up to 9.5 cm (3¾ in).
Distribution
Western Indian Ocean and tropical eastern Pacific, including Gulf of California.
Food
In nature: plankton floating on the bottom and small fish. In aquarium: does well on live *Artemia* mixed with any kind of dried fish food it will take.
General care
It is a sturdy fish which lives peacefully with other species. Create small crevices and install corals for protection. Keep in an aquarium of 80 liters (17½ gal) or more, at temperatures of 20° – 27°C (68° – 80°F).
Notes
This species is generally found in the upper part of coral formations, hiding in the branches. The red spots that adorn its body change tone according to the surroundings. They may, in fact, become pink, orange or red depending on the type of coral on which the fish happens to light. It adapts well to aquarium life, as do all others of the same genus, provided there are sufficient cracks and crevices in the rocks.

Chromis limbaughi

Chromis scottii

Cirrhitichthys oxycephalus

Chrysiptera cyanea

82 • TROPICAL SEA WATERS

CORIS GAIMARDI
Fam. Labridae

Common name
Clown wrasse.
Dimensions
Up to 20 cm (8 in).
Distribution
Tropical Indo-Pacific region; absent in Hawaii.
Food
In nature: small benthic animals. In aquarium: *Artemia*, ground and flaked fish food, krill, etc.
General care
These are tranquil fish. Aquarists tend to prefer the brightly coloured young. As with other similar species, the bottom of the tank should be covered with well aerated sand. Maintain temperatures of 22° – 27°C (72° – 80°F).
Notes
As often occurs among the Labridae, there is a marked colour difference between young and adults. The young of *C. gaimardi*, in fact, are red with white, black-edged spots which gradually disperse, vanishing in the adults who are violet with numerous blue specks on the flanks and with broad blue and green horizontal stripes on the opercula. In the aquarium it is best to keep only young specimens which, unlike the adults, can live with invertebrates.

CORYPHOPTERUS PERSONATUS
Fam. Gobiidae

Common name
Masked goby.
Dimensions
Up to 4 cm (1½ in).
Distribution
From Bermuda to the Virgin Islands.
Food
In nature: principally plankton. In aquarium: *Artemia* nauplii, live plankton (if possible), dried fish food.
General care
Keep in a tank without other species. Prepare cavities as hiding places. Recommended temperatures are 22° – 28°C (72° – 82°F).
Notes
This is one of the most widespread species of the tropical western Atlantic. It may be distinguished quite easily from other Gobiidae of the area (though not from *C. hyalinus*, whose identification depends on the different number of pores on the head – 2 rather than 3) by its orange colour, black head and small white spots and rectangular markings along the lateral line. The second ray of the dorsal fin is markedly longer than the others. Unlike most of the Gobiidae, who remain on the substrate, *C. personatus* swims in open water, often living in shoals around coral where it sometimes shelters in cavities. It is found at depths of 3 – 40 m (10 – 135 ft).

CROMILEPTES ALTIVELIS
Fam. Serranidae

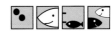

Common name
Leopard grouper.
Dimensions
Up to 65 cm (26 in).
Distribution
Indo-Pacific region from India to Philippines.
Food
In nature: carnivorous. In aquarium: feeds on pieces of fish, squid, shrimp, bivalves and almost any other edible sea creatures.
General care
Keep in an aquarium with other species. Prepare an environment similar to a coral reef with holes and crevices among the rocks. Use as large a tank as possible and maintain temperatures at 22° – 28°C (72° – 82°F).
Notes
This is one of the most easily obtainable members of the family for the aquarium. Although adults can reach a length of 65 cm (26 in), the young grow slowly in captivity, are tranquil and can be kept for several years in the home. In the wild they are benthic predators, generally staying partly hidden among rocks and coral, awaiting their prey.

DASCYLLUS ARUANUS
Fam. Pomacentridae

Common name
Three-striped humbug.
Dimensions
Up to 6 cm (2½ in).
Distribution
Widespread in tropical Indo-Pacific region.
Food
In nature: feeds on small crustaceans. In aquarium: after a while it may become accustomed to eating dead and dried food of a size commensurate with its natural prey.
General care
It can be reared in small shoals provided the aquarium is sufficiently roomy and offers plenty of hiding places. Recommended temperatures are 25° – 27°C (77° – 80°F).
Notes
As a rule, it lives around coral formations in lagoons and off the coast, from the surface to a depth of 12 m (40 ft). The conspicuous black and white coloration helps the members of the shoal to recognize one another.

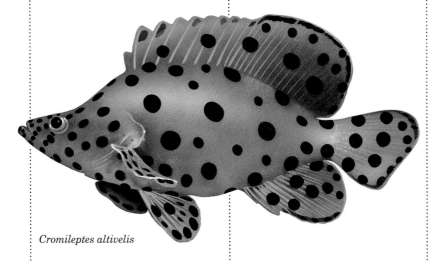

Cromileptes altivelis

Dascyllus aruanus

Coryphopterus personatus

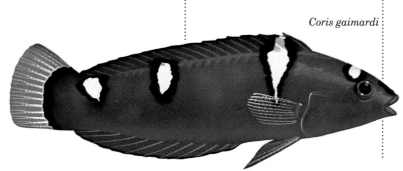

Coris gaimardi

TROPICAL SEA WATERS

DASCYLLUS MELANURUS
Fam. Pomacentridae

Common name
Black-tailed humbug.
Dimensions
Up to 6.5 cm (2¾ in).
Distribution
Principally tropical regions of the western Pacific.
Food
In aquarium: at first it can be fed on live food but it gradually becomes accustomed to dead and dried food.
General care
Several individuals can easily be kept together. A tranquil fish, it will live happily with other species and invertebrates.
Notes
This species lives in particularly calm and sheltered parts of the lagoon. It breeds quite easily in captivity.

DASCYLLUS TRIMACULATUS
Fam. Pomacentridae

Common name
Domino damsel, three-spot damselfish.
Dimensions
Up to 15 cm (6 in).
Distribution
Widely found in tropical Indo-Pacific region, from surface to a depth of 55 m (180 ft).
Food
Similar to other *Dascyllus* species, but also feeds on algae.
General care
An easy species to rear which may even breed given a favourable environment, laying eggs on the rocks in the aquarium.
Notes
The young are quite placid whereas the adults are aggressive among themselves and towards other species. The white spots to which the species owes its common name are conspicuous in the young but tend to disappear in the adults.

DENDROCHIRUS ZEBRA
Fam. Scorpaenidae

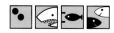

Common name
Zebra scorpion fish.
Dimensions
Up to 18 cm (7 in).
Distribution
Widespread in tropical Indo-Pacific region; absent in Hawaii.
Food
In nature: carnivorous. In aquarium: pieces of fish, squid and shrimp; if necessary begin with live food.
General care
As for *Pterois volitans*, but a 75-liter (16½-gal) aquarium is large enough.
Notes
This species, widely found around coral reefs, is easily caught, although care must be taken to avoid its poisonous spines. In Hawaii there is *D. barberi* (= *D. brachypterus*), a close relative, which seems to prefer caves and cliffs. Other similar species kept in the aquarium include *Pterois sphex* from Hawaii, *Pteropterus* (= *Pterois*) *antennata,* and *Pterois russelli* (= *P. lunulata*) from the western part of the Indo-Pacific region.

DIODON HOLACANTHUS
Fam. Diodontidae

Common name
Long-spined porcupine fish.
Dimensions
Up to 45 cm (18 in).
Distribution
Circumtropical.
Food
In nature: invertebrates. In aquarium: chopped seafood.
General care
Small specimens can be kept with other species. Maintain temperatures of 20° – 28°C (68° – 82°F).
Notes
This species is not usually selected by most aquarists because of its size. However, it is easily obtainable and caught throughout the tropical zones, like its relative *Diodon histrix*.

Dascyllus melanurus

Dascyllus trimaculatus

Dunkerocampus dactyliophorus

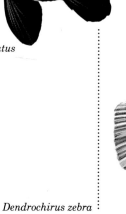

Dendrochirus zebra

Diodon holacanthus

DUNKEROCAMPUS DACTY-LIOPHORUS Fam. Syngnathidae

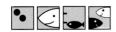

Common name
Banded pipefish.
Dimensions
Up to 18 cm (7 in).
Distribution
Tropical central Indo-Pacific region, including Indonesia and Melanesia.
Food
In nature: carnivorous; crustaceans living on seabed, larvae, fish, etc. In aquarium: live *Artemia*, newborn guppies, small fish.
General care
Adapts well to living in tanks with other species provided they are not aggressive. Take great care to ensure they are fed adequately. Maintain temperatures of 22° – 28°C (72° – 82°F).
Notes
This species, usually considered by reason of its coloration to be the loveliest of the Syngnathidae, is one of the few in this family to be of interest to the aquarist. It is found along coral reefs and can be caught with a narrow-mesh net. The male cares for the eggs, carrying them on his abdomen.

EMBLEMARIA HYPACANTHUS Fam. Chaenopsidae

Common name
Banner blenny.
Dimensions
Up to 5 cm (2 in).
Distribution
Gulf of California.
Food
In nature: crustaceans living on the seabed. In aquarium: *Artemia* nauplii, other live planktonic organisms, small worms; mix with dried fish food.
General care
Territorial, so it is advisable to keep them in an aquarium without other species. Try one female with two or three males in a 75-liter (16½-gal) tank. Introduce rocks containing shells, crevices or holes. Maintain temperatures of 20° – 23°C (68° – 74°F).
Notes
Males of *E. hypacanthus* live in tubular holes and are highly territorial. They announce their presence by waving their enlarged sail-like dorsal fin while at the entrance of their holes. Females, lacking the large fin, can be found both inside and outside the holes. The male tries to attract the female by entering and leaving the hole rapidly, waving his dorsal fin. If the female is receptive, she enters to copulate and lay her eggs, then swims off, leaving the male to watch over the eggs until they hatch.

EPINEPHELUS DERMATOLEPIS Fam. Serranidae

Common name
Leather grouper.
Dimensions
Up to 1 m (3 ft).
Distribution
From Gulf of California to Ecuador, including Galapagos Islands and other islands far from shore.
Food
In nature: carnivorous. In aquarium: takes chopped fish, prawns, squid, etc.
General care
Only the young are suitable for domestic aquaria. Despite their carnivorous nature, they are peaceful in a tank, living contentedly with other species, so long as the latter are not small enough to be eaten. Use a large aquarium and maintain temperatures at 18° – 24°C (65° – 75°F).
Notes
The young conceal themselves among the spines of the sea urchins *Diadema* and *Centrostephanus*, the black stripes rendering them virtually invisible. They can easily be caught in shallow water by scuba divers with or without cylinders, with a normal net. When they grow too large for home tanks, they can be offered to public aquaria.

EQUETUS LANCEOLATUS Fam. Sciaenidae

Common name
Jackknife fish.
Dimensions
Up to 23 cm (9 in).
Distribution
Tropical western Atlantic from Bermuda to Brazil.
Food
In nature: omnivorous, nocturnal. In aquarium: its diet is usually composed of live food such as *Artemia*; mix with krill, chopped shrimp or plankton.
General care
Prepare crevices or introduce coral so that they can hide. They adapt best to aquaria reserved for them alone but can be kept with others, provided the latter are very peaceful. Maintain temperatures of 22° – 24°C (72° – 75°F).
Notes
This species normally lives at depths of around 15 m (50 ft). *E. acuminatus* is more widespread and familiar to aquarists, but less attractive. Both species are nocturnal.

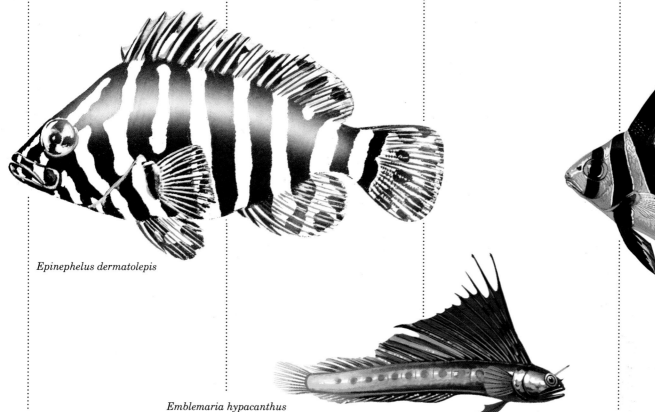

Epinephelus dermatolepis

Emblemaria hypacanthus

Equetus lanceolatus

EUXIPHIPOPS NAVARCHUS
Fam. Pomacanthidae

Common name
Blue-girdled angelfish, majestic angelfish.
Dimensions
Up to 25 cm (10 in).
Distribution
Indo-Australian archipelago, including Philippines.
Food
In nature: omnivorous. In aquarium: normally accepts a vast range of food including flaked fish food.
General care
Adapts well to aquarium life with other peaceful species. Provide many dimly lit hiding places and maintain a temperature of 22° – 28°C (72° – 82°F).
Notes
Young of around 8 cm (3 in) are often available for home aquaria. Keep one or two individuals in a large tank with other species. Sometimes they are timid at first, but once adapted they are worth the effort. The species can be found in sheltered lagoons and open zones with coral formations. As a rule these fish are solitary.

EXALLIAS BREVIS
Fam. Blenniidae

Common name
Spotted rock blenny.
Dimensions
Up to 15 cm (6 in).
Distribution
Tropical Indo-Pacific region from Sri Lanka to central Polynesia and Hawaii.
Food
In nature: omnivorous. In aquarium: live *Artemia*, krill and chopped seafood; mix with dried fish food containing vegetable matter.
General care
It is rarely kept in an aquarium with other species because it rips the fins of its tanks companions. Provide hiding places and maintain temperatures of 22° – 28°C (72° – 82°F).
Notes
The habitat of this species ranges from the shallow part of the reef to depths of at least 109 m (33 ft). Aquarists can choose from many Blenniidae species when filling their tank, but rest assured that if you notice the fins of your Pomacentridae and Pomacanthidae are damaged, this species is the culprit.

FORCIPIGER FLAVISSIMUS
Fam. Chaetodontidae

Common name
Long-nosed butterfly fish.
Dimensions
Up to 15 cm (6 in).
Distribution
Indian and Pacific Oceans.
Food
In nature: omnivorous, feeding on live invertebrates in cracks and crevices. In aquarium: adapts to a wide variety of food, including chopped fish, shrimp and squid, *Artemia* and flaked fish food.
General care
Lives happily in an aquarium of 300 liters (66 gal) or more, at temperatures of 21° – 28°C (70° – 82°F).
Notes
This lovely species is relatively easy to keep provided it is not disturbed by its aquarium companions. It feeds on live organisms in the crevices of coral reefs, stony and rock walls, to depths of over 30 m (100 ft). A close relation, *F. longirostris*, has a longer snout but is not often available to the aquarist.

GIBBONSIA ELEGANS
Fam. Clinidae

Common name
Spotted kelpfish.
Dimensions
Up to 16 cm (6¼ in).
Distribution
From central California to Baia Magdalena (Mexico); Guadelupe.
Food
In nature: carnivorous. In aquarium: chopped seafood.
General care
Suitable for home aquaria at temperatures of 23°C (74°F). Introduce rocks and vegetation (including artificial) as shelters. Do not keep with aggressive species.
Notes
This fish likes warm water and has certain characteristics in common with *Labrisomus xanti*. It lives generally in rock pools, concealing itself almost perfectly among the vegetation. It can be caught quite easily with a hand net after draining pools. The female lays eggs among the plants and the male cares for them until they hatch. Many other species, closely related to *G. elegans*, live in the same geographical zone.

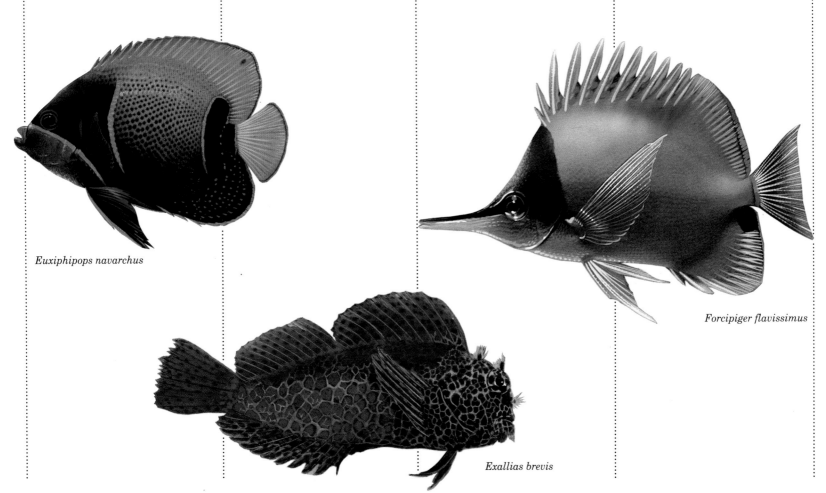

Euxiphipops navarchus

Forcipiger flavissimus

Exallias brevis

GOBIODON CITRINUS
Fam. Gobiidae

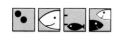

Common name
Lemon goby.
Dimensions
Up to 6 cm (2½ in).
Distribution
Tropical Indo-Pacific region from East Africa to Fiji.
Food
In nature: carnivorous; worms and other invertebrates, fish eggs. In aquarium: *Artemia* nauplii, small worms, fish eggs, dried fish food, frozen krill.
General care
Arrange some coral branches as shelters. The fish can live in tanks with more than one species but it is better if they have a tank of their own. Temperatures should be 22° – 28°C (72° – 82°F).
Notes
This species lives on coral and lays eggs in the angles of basal branches. It displays a vast range of colours that run from greenish to yellow and bright orange.

GOBIOSOMA DIGUETI
Fam. Gobiidae

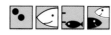

Common name
Banded cleaner goby.
Dimensions
Up to 3 cm (1¼ in).
Distribution
From Gulf of California to Colombia.
Food
In nature: behaves like a cleaner fish but also feeds on invertebrates that live at the surface. In aquarium: *Artemia* nauplii, surface copepods, small worms, dried fish food.
General care
This is a very small fish which needs safe refuge places, such as holes and cracks. It can be kept with other species. Temperatures should be 20° – 23°C (68° – 74°F).
Notes
This fish can be seen ridding large predatory fish, such as Serranidae and Muraenidae, of parasites. It would appear to possess a noxious mucus that protects it from such predators.

GOBIOSOMA OCEANOPS
Fam. Gobiidae

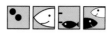

Common name
Neon goby.
Dimensions
Up to 6 cm (2½ in).
Distribution
Florida and Yucatan.
Food
In nature: it is a cleaner fish which also feeds on small benthic crustaceans. In aquarium: voracious eater of normal dried fish food.
General care
It is an excellent aquarium species, strong and active, yet peaceful. It is suitable for keeping with other species except those of the genus *Hypoplectrus*, which will eat it. Make refuges with stones and coral. Maintain temperatures of 22° – 28°C (72° – 82°F).
Notes
This is one of the many species of Gobiidae which behaves like a cleaner fish. Its colours and markings are very similar to those of *Labroides dimidiatus* from the Indo-Pacific zone. *G. oceanops* lives down to a depth of 40 m (135 ft), usually among branches of coral but also in caves and crevices. The eggs are laid in a hole, watched over by the male until they hatch, normally within 8 – 12 days. It has been bred in captivity.

GOBIOSOMA PUNCTICULATUS
Fam. Gobiidae

Common name
Red-headed goby.
Dimensions
Up to 5 cm (2 in).
Distribution
From Gulf of California to Ecuador.
Food
In nature: carnivorous; sometimes behaves like a cleaner fish. In aquarium: *Artemia* nauplii, small worms and crustaceans, mixed with dried fish food.
General care
It can be kept in a tank with other species provided it has adequate places of refuge, such as holes, cracks and crevices. Temperatures should be 20° – 23°C (68° – 74°F).
Notes
This is a territorial species which often lives commensally with sea urchins. Territorial disputes among individuals of *G. puncticulatus*, involving bites, often lead to severe injuries or even death.

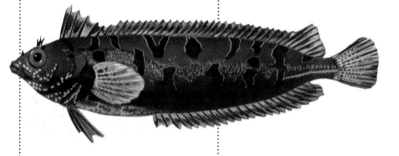

Gibbonsia elegans

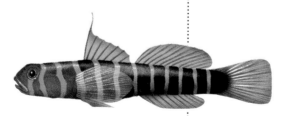

Gobiosoma digueti

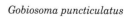

Gobiosoma oceanops

Gobiodon citrinus

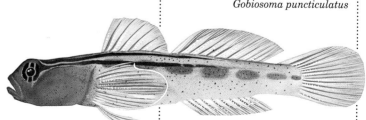

Gobiosoma puncticulatus

GOMPHOSUS VARIUS
Fam. Labridae

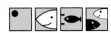

Common name
Bird wrasse.
Dimensions
Up to 25 cm (10 in).
Distribution
Tropical Indo-Pacific region, including Hawaii.
Food
In nature: principally shrimp. In aquarium: chopped seafood mixed with vegetable matter.
General care
Keep in an aquarium of 200 liters (44 gal) or more, with a well aerated sandy bed. Do not introduce invertebrates to the tank and create plenty of hiding places, which are indispensable if keeping pairs, so as to allow the female to escape from the continual pursuit of the male.
Notes
The colour of the males varies from green to blue–green; the females are brown. This contrast in coloration sometimes causes the two sexes to be described under different names. In some books the species is classified as *Gomphosus coerulens*. It is a popular and robust aquarium fish, adapting well to life with other species.

GRAMMA LORETO
Fam. Grammidae

Common name
Royal gramma.
Dimensions
Up to 8 cm (3 in).
Distribution
Tropical western Atlantic from Bermuda to Venezuela, but absent in Florida.
Food
In nature: plankton. In aquarium: live *Artemia*, krill and other small crustaceans, dried fish food.
General care
It lives best in an aquarium reserved for this species. Several individuals can be kept together provided each has its own "cave" surrounded by a territory. Maintain temperatures of 24° – 27°C (75° – 80°F).
Notes
G. Loreto is found mainly around reefs at depths of 12 – 24 m (40 – 80 ft), but may also be seen at only 3 m (10 ft) or so. Individuals that live in shallow water usually inhabit caves, where they can be observed swimming upside down near the surface.

GRAMMISTES SEXLINEATUS
Fam. Grammistidae

Common name
Golden-striped grouper.
Dimensions
Up to 30 cm (12 in).
Distribution
Tropical Indo-Pacific region from East Africa to Tahiti; not present in Hawaii.
Food
In nature: carnivorous. In aquarium: fish, squid, lean ox heart.
General care
This is a difficult fish to keep because if it is handled or frightened, it gives out a toxic mucus that can even kill other aquarium occupants. It should never be transported with other fish. If you decide to have *G. sexlineatus*, try to keep it on its own in a large tank where it can remain undisturbed until acclimated. Then introduce other fish, provided they are bigger. Maintain temperatures of 22° – 27°C (72° – 80°F).
Notes
It lives on coral reefs and in rock pools. It is very timid and, if approached, darts rapidly into shelter. In the aquarium, however, it may become quite tame.

GYMNOMURAENA ZEBRA
Fam. Muraenidae

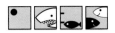

Common name
Zebra moray.
Dimensions
Up to 76 cm (30 in).
Distribution
Pacific and western Indian Ocean.
Food
In nature: small benthic invertebrates, especially crustaceans. In aquarium: adapts to chopped shrimp, squid, etc.
General care
A shy species. Provide shelters such as crevices in rocks. Often stays hidden by day, except when feeding. Recommended temperatures are 17° – 30°C (63° – 86°F).
Notes
Lives among reefs. Unlike the majority of moray eels, *G. zebra* has rounded teeth similar to molars, suited to cracking small hard-shelled organisms. It is most active at night and scuba divers rarely catch a glimpse of it in daytime.

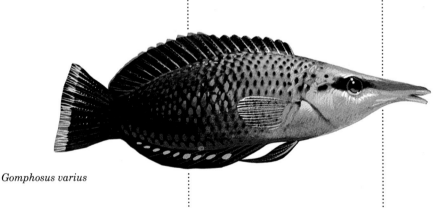

Gomphosus varius

Gramma loreto

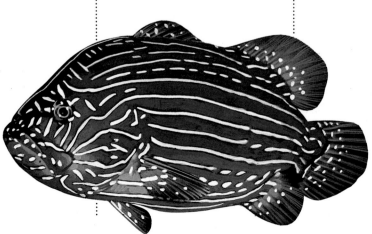

Gymnomuraena zebra

Grammistes sexlineatus

GYMNOTHORAX MELEAGRIS
Fam. Muraenidae

Common name
Spotted moray.
Dimensions
Up to 100 cm (40 in).
Distribution
Tropical zones of Indian and Pacific Oceans from Africa to Polynesia.
Food
In nature: carnivorous. In aquarium: chopped fish, squid and shrimp.
General care
Suitable for keeping in aquaria with other species, apart from those on which it feeds. Create crevices for hiding. Maintain temperatures of 22° – 28°C (72° – 82°F).
Notes
Like many moray eels, *G. meleagris* is a nocturnal predator, relying greatly on its sense of smell to locate prey. It is very common along coral reefs where it may be caught quite easily, but it struggles to get free once it has taken the bait. With a little patience, it can also be caught on the high seas, even with hand nets. Smaller specimens are very interesting for home aquaria, which should be of 75 liters (16½ gal) or more. This eel is hardy and disease-resistant.

HAEMULON SEXFASCIATUM
Fam. Haemulidae

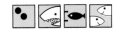

Common name
Graybar grunt.
Dimensions
Up to 30 cm (12 in).
Distribution
From Gulf of California to Panama.
Food
In nature: carnivorous, benthic animals, including small fish, crustaceans, molluscs and worms. In aquarium: chopped molluscs.
General care
The young can live in tanks of 75 liters (16½ gal) or more. As a rule, the adults are for display only in public aquaria. They are long-lived in captivity. Keep at temperatures of 20° – 23°C (68° – 74°F).
Notes
By day they gather close to reefs, dispersing to feed at night, the time when they can be caught with trawl nets in rocky zones and off pebble beaches.

HEMIPTERONOTUS PAVONINUS
Fam. Labridae

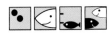

Common name
Razorfish.
Dimensions
Up to 25 cm (10 in).
Distribution
Tropical Pacific and Indian Oceans.
Food
In nature: small invertebrates and fish, including the Congridae. In aquarium: chopped seafood.
General care
It needs a bed covered with at least 7 cm (2¾ in) of well aerated sand. The adults must be given an aquarium of at least 1,200 liters (264 gal).
Notes
This species is common in sandy zones close to coral reefs. If threatened, or at night, it hides in the sand. It can be caught with rod and line and small hooks, or with traps. It is a highly aggressive species.

HEMITAURICHTHYS POLYLEPIS
Fam. Chaetodontidae

Common name
Pyramid butterfly fish.
Dimensions
Up to 18 cm (7 in).
Distribution
East Indies, central and western Pacific, including Hawaii and Great Barrier Reef.
Food
In nature: principally plankton. In aquarium: chopped shrimp, *Artemia* and other crustaceans, flaked fish food.
General care
Prefers to live together with other individuals of its own kind in an average tank of 300 liters (66 gal) or more. Maintain temperatures of 24° – 28°C (75° – 82°F).
Notes
A gregarious species which is normally found at depths of 5 – 25 m (15 – 80 ft), feeding principally on plankton. For reasons unknown, perhaps stress and lighting, the coloration of the head may vanish in captive specimens. In the Indian Ocean there is a very similar species, *H. zoster*, which is dark brown.

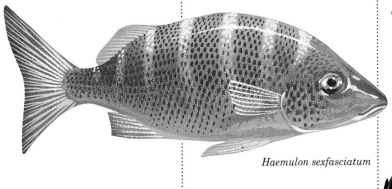

Haemulon sexfasciatum

Hemipteronotus pavoninus

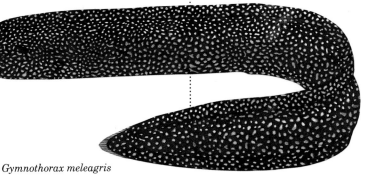

Hemitaurichthys polylepis

Gymnothorax meleagris

TROPICAL SEA WATERS • 89

HENIOCHUS ACUMINATUS
Fam. Chaetodontidae

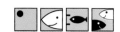

Common name
Pennant coralfish, wimplefish.
Dimensions
Up to 25 cm (10 in).
Distribution
Pacific and Indian Oceans, including Hawaii, Japan, Indonesia, Australia, South Africa and Red Sea.
Food
In nature: omnivorous. In aquarium: principally dried fish food, including flaked forms; it is as well to vary the diet.
General care
Recommended for average or large aquaria, with temperatures of 21°–28°C (70°–82°F). It prefers to live with others of its own species.
Notes
This species exhibits considerable variations in its geographical range and may give the impression of being a collection of different though similar species. It is normally found at the surface of coral reefs, alone, in pairs or in small groups. Large shoals of a similar species, *H. diphreutes*, are often seen in mid water. It usually lives below 12 m (40 ft) down to a depth of 150 m (500 ft). It is easily caught and is a particularly suitable species for aquaria.

HIPPOCAMPUS KUDA
Fam. Syngnathidae

Common name
Yellow seahorse.
Dimensions
Up to 25 cm (10 in).
Distribution
Tropical Indo-Pacific region.
Food
In nature: zooplankton. In aquarium: offer newly hatched nauplii and live adult *Artemia*. Try to mix with other live crustaceans and small fish such as newborn guppies.
General care
Keep with other tranquil species, such as needlefish, or alone with their own kind. Provide vegetation or coral to which they can attach themselves.
Notes
The marine seahorses are normally among the most difficult species to keep in the aquarium because they need a varied and plentiful diet, including mainly live food. Coloration is a good sign of the animal's state of health: a yellow livery is typical of healthy fish, turning progressively darker until it becomes nearly black in sick and distressed individuals.

HISTRIO HISTRIO
Fam. Antennariidae

Common name
Sargassum fish.
Dimensions
Up to 20 cm (8 in).
Distribution
Tropical Atlantic and western Pacific among drifting sargassum weed.
Food
In nature: carnivorous; small fish. In aquarium: start with small live fish, then go on to chopped fish, shrimp and squid when adapted to captive conditions.
General care
If possible, install floating sargassum weed for concealment, otherwise an imitation of the plant. Maintain temperatures of 20°–27°C (68°–80°F).
Notes
This highly cryptic fish is coloured and shaped like sargassum weed; it rarely abandons this refuge, where it preys on many small fish.

HOLACANTHUS BERMUDENSIS
Fam. Pomacanthidae

Common name
Blue angelfish.
Dimensions
Up to 45 cm (18 in).
Distribution
Western Atlantic, Bermuda and Gulf of Mexico; occasionally in Bahamas.
Food
In nature: sponges, tunicates and other invertebrates. In aquarium: as a rule it adapts quickly to a diet of various forms of seafood, mixed with flaked fish food.
General care
As for *H. ciliaris*.
Notes
The species, known also as *H. isabelita*, exhibits a marked diversity of colour from the young to adult stage. The former are bluish with vertical blue stripes, while the snout, tail and ventral and pectoral fins are orange–yellow. The adults are blue–green with purple tints and yellow-bordered caudal and pectoral fins. In the waters of Florida and the Bahamas, where the species lives, crosses with *H. ciliaris* have been observed.

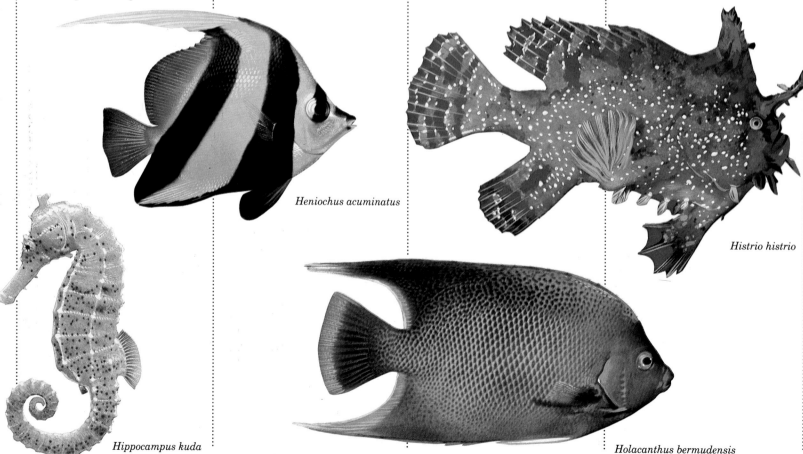

Heniochus acuminatus

Histrio histrio

Hippocampus kuda

Holacanthus bermudensis

HOLACANTHUS CILIARIS
Fam. Pomacanthidae

Common name
Queen angelfish.
Dimensions
Up to 45 cm (18 in).
Distribution
Tropical western Atlantic from Brazil to Florida.
Food
In nature: sponges and other sessile invertebrates; some algae. In aquarium: fresh or frozen seafood, flaked fish food.
General care
It is a robust species. Adults require a large aquarium of 500 liters (110 gal) or more, maintained at temperatures of 20° – 28°C (68° – 82°F).
Notes
The species is distinguished from its close relative *H. bermudensis* by having a black growth, covered with blue spots, on its forehead. It is possible to find hybrids of two species, commonly known as Townsend's angel, when their distribution zones overlap. Specimens of *H. ciliaris* live alone or in pairs on the coral reef, from the surface to a depth of 70 m (230 ft). The young sometimes act as cleaners, removing ectoparasites from other fish.

HOLACANTHUS CLARIONENSIS
Fam. Pomacanthidae

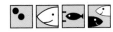

Common name
Clarion angelfish.
Dimensions
Up to 25 cm (10 in).
Distribution
Eastern Pacific, including Revillagigedo Islands (Mexico), extreme south of Lower California and Guadelupe.
Food
In nature: omnivorous, feeding on a wide variety of plants and invertebrates. In aquarium: adapts to vegetable and animal based foods.
General care
Keep the adults in a large aquarium of 400 liters (88 gal) or more at temperatures of 20° – 28°C (68° – 82°F).
Notes
These vividly coloured fish are very common in the Revillagigedo Islands, 400 km (250 miles) south of Cape San Luca (Mexico), where they gather at sunset in large shoals to feed; at this time they are particularly easy to catch with hand nets. The adults range from the surface down to a depth of at least 70 m (230 ft). The young have colours similar to those of the young of *H. passer* and normally live in shallow water.

HOLACANTHUS PASSER
Fam. Pomacanthidae

Common name
King angelfish.
Dimensions
Up to 30 cm (12 in).
Distribution
Eastern Pacific from Gulf of California to Galapagos Islands.
Food
In nature: omnivorous, including sponges and plants. In aquarium: fresh or frozen seafood and vegetable matter.
General care
The young can be kept in small aquaria but those of at least 150 liters (33 gal) are recommended. The adults need large tanks. To reduce incidence of fights, choose individuals that differ from one another in length by at least 3 cm (1¼ in). They live happily at temperatures of 20° – 24°C (68° – 75°F).
Notes
These fish live from the surface down to a depth of at least 80 m (265 ft). A scuba diver can easily catch them during the night.

HOLOCENTRUS RUFUS
Fam. Holocentridae

Common name
White-tip squirrelfish.
Dimensions
Up to 20 cm (8 in).
Distribution
From Bermuda to Lesser Antilles.
Food
In nature: carnivorous, principally crustaceans. In aquarium: a diet consisting of chopped fish, shrimp and squid.
General care
Easy to raise in an aquarium with one or more species. It is best to keep several individuals together, in a tank of at least 350 liters (77 gal) in which caves have been created. Maintain temperatures of 18° – 25°C (65° – 77°F).
Notes
During the day this species lives in dimly lit zones, beneath overhangs or in caves, at a depth of 30 m (100 ft).

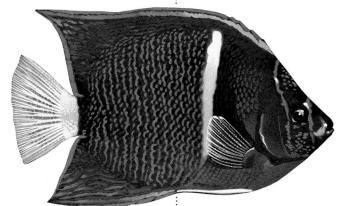

Holacanthus passer

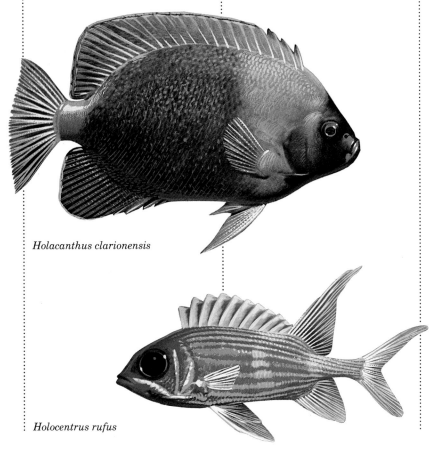

Holacanthus clarionensis

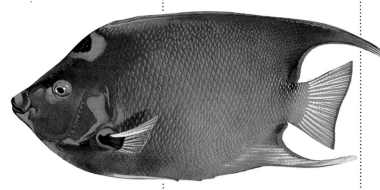

Holacanthus ciliaris

Holocentrus rufus

TROPICAL SEA WATERS

HYPOPLECTRUS UNICOLOR
Fam. Serranidae

Common name
Hamlet.
Dimensions
Up to 13 cm (5¼ in).
Distribution
From Florida to Central America.
Food
In nature: carnivorous, principally shrimp, crabs and small fish. In aquarium: chopped shrimp and small fish.
General care
These fish are happy to live with other species in an aquarium of 75 liters (16½ gal) or more. The young, however, adapt better to captivity. Provide many hiding places and maintain temperatures of 21° – 25°C (70° – 77°F).
Notes
This species was once classified as three separate species which differed from one another only in colour. Following recent biochemical studies, *H. gemma*, *H. gummigata* and *H. guttavarius* are now recognized as varieties of the single species, *H. unicolor*. Each of the above-mentioned varieties appears to imitate a different species (*H. cyaneus*, *H. tricolor*), living peacefully alongside them.

HYPSOBLENNIUS GENTILIS
Fam. Blenniidae

Common name
Bay blenny.
Dimensions
Up to 15 cm (6 in).
Distribution
From central California to Baia Magdalena (Mexico), northern and central Gulf of California.
Food
In nature: benthic invertebrates and algae. In aquarium: chopped seafood, mixed with krill, live *Artemia* and, if desired, dried fish food.
General care
May be aggressive towards sedentary species. Tolerates a vast range of temperatures and does not require a heated aquarium. Introduce stones and shells as possible hiding places.
Notes
This fish easily becomes tame in captivity. It is common in rock pools in the northern part of the Gulf of California and in estuaries along the outer coastline of California and Lower California. In rock pools along the open coasts, *H. gentilis* is replaced by *H. gilberti*, whereas *H. jenkinsi* lives mainly in caves, on mussel beds and around posts. This last species does not adapt to aquarium life as easily as *H. gentilis*.

HYPSYPOPS RUBICUNDUS
Fam. Pomacentridae

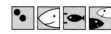

Common name
Garibaldi damselfish.
Dimensions
Up to 35 cm (14 in).
Distribution
From Monterey Bay to Baia Magdalena (Mexico), Lower California, Mexico.
Food
In nature: omnivorous, but normally eats a large quantity of sponges. In aquarium: adapts to a diet of fish and squid but accepts and thrives well on dried fish food.
General care
The adults are big, highly territorial and, in community tanks, constantly nibble at sedentary fish. A single adult needs a tank of 750 liters (165 gal) or more. If various individuals are kept, create a crevice or hiding place for each of them. Maintain temperatures of 13° – 24°C (55° – 75°F).
Notes
The young are brightly coloured and can normally be seen in shallow water among iridescent red algae in July–August. Considering their vivid coloration, they are not all that conspicuous in their surroundings. The juveniles are a fairly drab orange with blue markings and grow to about 10 cm (4 in).

JOHNRANDALLIA NIGRIROSTRIS
Fam. Chaetodontidae

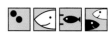

Common name
Blacknosed butterfly fish.
Dimensions
Adults up to about 20 cm (8 in).
Distribution
From Gulf of California to Gulf of Panama and Galapagos Islands.
Food
In nature: encrustant organisms or organisms floating on the bottom, crustaceans, gastropods and some algae. In aquarium: feed with a mixed diet of shrimp and squid, mixed with *Artemia* and flaked fish food.
General care
Keep at temperatures of 20° – 24°C (68° – 75°F).
Notes
Some individuals act as cleaner fish or "cleaning stations," where larger fish (hosts) gather to receive their services. Usually found from the surface to a depth of 12 m (40 ft), these fish also venture down to at least 40 m (135 ft).
They live alone, in pairs (often observed at night) or in shoals. This species can easily be caught at night by scuba divers with torches and nets.

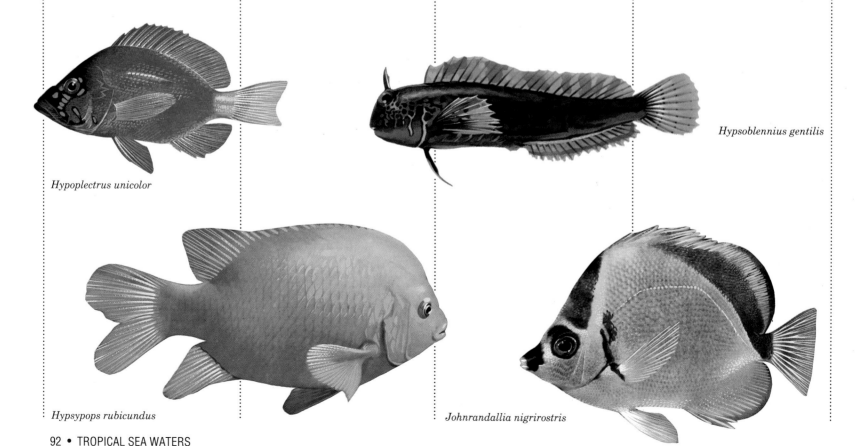

Hypoplectrus unicolor

Hypsoblennius gentilis

Hypsypops rubicundus

Johnrandallia nigrirostris

KUHLIA TAENIURA
Fam. Kuhliidae

Common name
Ahole hole.
Dimensions
Up to 30 cm (12 in).
Distribution
Tropical zones of Indo-Pacific, from East Africa to Revillagigedo Islands (Mexico).
Food
In nature: feeds at night, principally on free-swimming crustaceans. In aquarium: accepts any kind of aquarium food, including fish, squid, bivalves and chopped shrimp.
General care
Adapts to life with others of its kind or with a number of other species. Keep several specimens together in a large tank with a sandy bed, provided with some shelters among the stones and corals. Make sure there is enough room for swimming. Lives happily at temperatures of 21° – 24°C (70° – 75°F).
Notes
The young of this species live in pools formed by the tide, where they are easily caught.

LABROIDES DIMIDIATUS
Fam. Labridae

Common name
Cleaner wrasse.
Dimensions
Up to 7 cm (2¾ in).
Distribution
Tropical Indo-Pacific region.
Food
In nature: it is a cleaner fish which also feeds on small free-swimming invertebrates. In aquarium: live *Artemia*, other planktonic organisms, chopped shrimp, flaked fish food.
General care
It is an attractive and useful fish which likes living together with other species. Maintain temperatures of 22° – 28°C (72° – 82°F).
Notes
When buying this fish, make sure not to purchase, by error, *Aspidontus taeniatus* (page 73), the false cleaner fish or sabre-toothed blenny, which has a similar outward appearance. The principal difference is in the position of the mouth: terminal in *L. dimidiatus*, ventral in *A. taeniatus*.

LACTORIA CORNUTA
Fam. Ostraciidae

Common name
Long-horned cowfish.
Dimensions
Up to 50 cm (20 in).
Distribution
Tropical Indo-Pacific region.
Food
At first accepts only live fresh food (*Artemia*, small fish, etc.); with considerable patience, and admittedly much difficulty, it can be induced to take food from the hand.
General care
Needs a sandy substrate, on which pieces of rock or branches of coral can be placed. Lighting should be good and the water temperature 24° – 30°C (75° – 86°F).
Notes
Do not keep this species together with restless or aggressive fish because it is placid and also very sensitive. Always on the move, it maneuvers like a tiny helicopter.

LACTORIA FORNASINA
Fam. Ostraciidae

Common name
Spiny cowfish.
Dimensions
Up to 15 cm (6 in).
Distribution
Tropical Indo-Pacific region, from East Africa to Hawaii.
Food
In nature: principally invertebrates living in sand. In aquarium: chopped seafood, worms, krill.
General care
This species secretes toxic mucus when agitated. Keep in a tank without other species or, if you want to make an aquarium with several associated species, wait a couple of weeks before adding other fish. Temperatures should be 22° – 28°C (72° – 82°F).
Notes
These fish should only be kept by experienced aquarists. Handle them with special care because they secrete toxin that can be fatal both to themselves and their aquarium companions. To transport them, pack them in separate bags.

Kuhlia taeniura

Lactoria cornuta

Labroides dimidiatus

Lactoria fornasina

TROPICAL SEA WATERS

LIENARDELLA FASCIATUS
Fam. Labridae

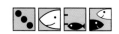

Common name
Harlequin wrasse.
Dimensions
Up to 25 cm (10 in).
Distribution
Tropical western Pacific, including islands of Ryukyu and Taiwan, Great Barrier Reef and Vanuatu.
Food
In nature: small invertebrates and fish. In aquarium: accepts an assortment of chopped fish, squid and molluscs.
General care
Recommended only to the expert aquarist. Cover the bottom with well aerated sand in which it can burrow. Maintain temperatures of 23° – 26°C (74° – 79°F). The adults need an aquarium of at least 2,000 liters (440 gal).

LIOPROPOMA FASCIATUS
Fam. Serranidae

Common name
Striped sea bass.
Dimensions
Up to 18 cm (7 in).
Distribution
Gulf of California.
Food
In nature: not known, perhaps bottom-dwelling organisms. In aquarium: chopped fish, prawns and squid, mixed with live adult *Artemia*, krill, etc.
General care
An excellent aquarium fish that adapts well to living with others. Provide refuge holes and crevices over a sandy bed. Maintain temperatures of 19° – 22°C (67° – 72°F).
Notes
This species has rarely been seen in the wild. It is most commonly found below the surface thermocline at a depth of more than 25 m (80 ft), close to the entrance of sandy-bottomed caves. Because it has a closed swim bladder, it must be degassed at the depth of capture with a hypodermic needle.

LIOPROPOMA RUBRE
Fam. Serranidae

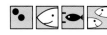

Common name
Swissguard basslet.
Dimensions
Up to 9 cm (3½ in).
Distribution
From Gulf of Mexico and southern Florida to Lesser Antilles; Yucatan.
Food
In nature: not known, perhaps small crustaceans. In aquarium: live *Artemia*, shrimp, squid, chopped bivalves, fish food.
General care
This fish is happy to live with other species. Provide caves and cracks as hiding places. Keep in an aquarium of 75 liters (16½ gal) or more at temperatures of 20° – 24°C (68° – 75°F).
Notes
In the wild it normally lives hidden in deep caves and cavities of the coral reef. As a rule, it is domesticated quite easily in the aquarium but if it continues to hide away, reduce the number of available shelters. It is not usually boisterous nor easily disturbed, except by more aggressive species.

LO VULPINUS
Fam. Siganidae

Common name
Foxface, foxfish.
Dimensions
Up to 25 cm (10 in).
Distribution
Tropical Pacific from East Indies to Marshall Islands.
Food
In nature: principally herbivorous. In aquarium: accepts any type of fish food mixed with vegetable matter.
General care
Aggressive towards individuals of its own species but peaceful with others. Keep one or a pair together with other varieties in a tank containing a spacious open zone and a sandy bed. The temperature should be 22° – 28°C (72° – 82°F).
Notes
This species is common in shallow water with plenty of seaweed, but it also lives in the vicinity of coral reefs and submerged wreckage. The spines of its dorsal fins are poisonous and may cause painful wounds. In fact, the fish, if disturbed or sensing itself threatened, usually raises these spines defensively, facing its assailants. At least two species of tropical Siganidae (*Siganus luridus* and *S. rivulatus*) have found their way into the Red Sea and the Mediterranean.

Liopropoma fasciatus

Liopropoma rubre

Lo vulpinus

Lienardella fasciatus

LUTJANUS SEBAE
Fam. Lutjanidae

Common name
Emperor snapper.
Dimensions
Up to 90 cm (36 in).
Distribution
Indo-Pacific region, except for Hawaii.
Food
In nature: carnivorous. In aquarium: chopped molluscs.
General care
Prepare a rocky environment with plenty of space for swimming and a sandy bed. The temperature should be 22° – 28°C (72° – 82°F).
Notes
Young individuals have brighter colours and proportionately larger fins, while the reddish brown bands tend to disappear almost completely in the adults. Once acclimated, these snappers take food without any problems, but care must be taken not to overfeed them as this would cause a dangerous build-up of fat in the internal organs. The species is prone to *Cryptocaryon irritans* infections.

LYTHRYPNUS DALLI
Fam. Gobiidae

Common name
Catalina goby, blue-banded goby.
Dimensions
Up to 6 cm (2½ in).
Distribution
From Central California to Gulf of California, Guadelupe.
Food
In nature: carnivorous, feeding principally on small crustaceans, including amphipods and copepods. In aquarium: live and frozen *Artemia*, chopped seafood, fish food.
General care
Prepare a rocky environment with plenty of holes and cracks. Keep in a tank reserved for this species or containing other small, tranquil species, at temperatures of 15° – 22°C (59° – 72°F). Several specimens could also be kept in a small 40-liter (8 ¾-gal) aquarium.
Notes
This species is found abundantly on reefs where there are plenty of cracks and holes. It lives at depths between the intertidal zone and 75 m (250 ft). It is short-lived, seldom surviving more than two years. The female lays eggs in summer, at the age of one year, in a concealed spot chosen by the male, who looks after them until they hatch.

MEIACANTHUS ATRODORSALIS
Fam. Blenniidae

Dimensions
Up to 7 cm (2¾ in).
Distribution
Tropical western Pacific from Moluccas to Samoa.
Food
In nature: carnivorous; planktonic crustaceans and benthic invertebrates. In aquarium: live *Artemia*, krill, squid, shrimp and chopped bivalves.
General care
It will live happily with other species. Prepare an aquarium of 75 liters (16½ gal) or more with crevices in which the fish can hide, and leave plenty of free space for swimming. The temperature should be 22° – 28°C (72° – 82°F).
Notes
This species lives in lagoons and out at sea on reefs, at depths of up to 30 m (100 ft) or more. It is difficult to catch and rarely available for sale. The canine teeth, grooved with poison sacs at the base, constitute protection against high-sea predators, which soon learn to leave it alone.

MELICHTHYS NIGER
Fam. Balistidae

Common name
Black triggerfish.
Dimensions
Up to 36 cm (14½ in).
Distribution
Tropical Pacific, Indian and Atlantic Oceans.
Food
In nature: omnivorous. It is capable of eating hard-shelled animals such as crabs and sea urchins. In aquarium: chopped fish, squid and bivalves, mixed with crab and vegetable matter.
General care
Small specimens are excellent for a home aquarium containing several associated species. They are hardy and active, but not aggressive. Maintain temperatures of 22° – 28°C (72° – 82°F).
Notes
The colour is dark green or black, with blue lines along the base of the second dorsal fin, which give this species a distinctive appearance. As a rule it lives on the outer slopes of the barrier reef. Like other Balistidae, it can be caught in the crevices by lowering the second spine of the dorsal fin to coax it out.

Lutjanus sebae

Lythrypnus dalli

Meiacanthus atrodorsalis

Melichthys niger

MICROSPATHODON DORSALIS
Fam. Pomacentridae

Common name
Giant damselfish.
Dimensions
Up to 30 cm (12 in).
Distribution
From Gulf of California to Colombia, Galapagos Islands.
Food
In nature: principally herbivorous. In aquarium: accepts shrimp, chopped fish, etc., mixed with vegetable matter.
General care
It is an aggressive species. Adults need a large aquarium of 1,200 liters (264 gal) or more, kept at temperatures of 20° – 28°C (68° – 82°F). Provide hiding places among round stones.
Notes
The young are distinguished by their conspicuous light blue spots along the back. These fish, including the young, sometimes bite their aquarium companions, but can be raised normally with other large and active species of, for example, the Pomacentridae and Chaetodontidae.

MONOCENTRIS JAPONICUS
Fam. Monocentrididae

Common name
Japanese pinecone fish.
Dimensions
Up to 13 cm (5¼ in).
Distribution
Widely found in the Indo-Pacific region.
Food
In nature: uncertain. In aquarium: normally takes whole or chopped shrimp and bivalves.
General care
Easy to keep; put several individuals together in an aquarium without other species. Prepare a dimly lit cave, with an open space in front for swimming. It adapts well at temperatures of 20° – 22°C (68° – 72°F).
Notes
These are among the rarest of aquarium fish. If available, they are to be recommended because they are sturdy, unaggressive, interesting and unique in appearance. Apart from *M. japonicus*, the Australian species *Cleidopus gloriaemaris* is sometimes available and equally desirable.

MULLOIDICHTHYS DENTATUS
Fam. Mullidae

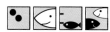

Common name
Goatfish.
Dimensions
Up to 30 cm (12 in).
Distribution
From southern California (rare) to Peru.
Food
In nature: feeds on small crustaceans, molluscs and worms. In aquarium: chopped shrimp, squid and bivalves.
General care
Smaller individuals are suitable for living with other species in tanks of 75 liters (16½ gal) or more; if possible use aquaria of larger capacity. Create ample sandy zones. The fish feed more readily in the evening. Keep temperatures at 20° – 23°C (68° – 74°F).
Notes
The fish form very compact shoals during the day and tend to disperse to feed at night, at which time they can be caught with a hand net or, more easily, with a fixed net. In the daytime *M. dentatus* often subjects itself to being "cleaned" by *Johnrandallia*. It shows itself ready for this operation by stationing itself head-downward and taking on a darker colour. A related species, *M. martinicus*, lives in the western Atlantic. Other species come from the Indo-Pacific region.

MURAENA PARDALIS
Fam. Muraenidae

Common name
Dragon moray.
Dimensions
Up to 100 cm (40 in).
Distribution
Indian and Pacific Oceans from southern Japan to Polynesia.
Food
In nature: carnivorous; prefers bottom-dwelling crustaceans and octopi, but also feeds on small fish. In aquarium: chopped seafood.
General care
Small individuals can live in aquaria of 75 liters (16½ gal) or more, with other species. It is a hardy fish, long-lived, and normally does not trouble its companions so long as they are not of a size to be eaten. Place rocks with crevices in the tank and maintain temperatures of 22° – 28°C (72° – 82°F).
Notes
Like the majority of moray eels, *Muraena pardalis* is active at night, but in captivity soon learns to feed by day. The horn-like growths above the eyes are actually tubular extensions to the posterior nostrils. A close relative, *M. lentiginosa*, is common in the Gulf of California and along the tropical eastern shores of the Pacific to Peru and the Galapagos Islands. It, too, is an excellent aquarium fish, although not readily available for the home aquarium.

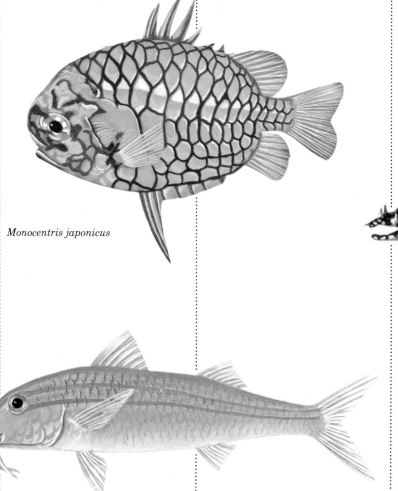

Monocentris japonicus

Mulloidichthys dentatus

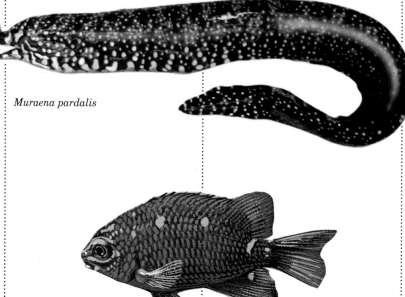

Muraena pardalis

Microspathodon dorsalis

MYRIPRISTIS MURDJAN Fam. Holocentridae	**NASO LITURATUS** Fam. Acanthuridae	**NASO UNICORNIS** Fam. Acanthuridae	**NEOCLINUS BLANCHARDI** Fam. Clinidae
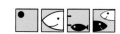	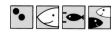		

MYRIPRISTIS MURDJAN — Fam. Holocentridae

Common name
Big-eye squirrelfish.
Dimensions
Up to 30 cm (12 in).
Distribution
Tropical Indo-Pacific region from East Africa to Tahiti.
Food
In nature: mainly crustaceans. In aquarium: fresh or frozen chopped squid, fish, shrimp, etc.
General care
Will live happily in tanks with other species. If possible, keep several individuals together. Create one or more dark rocky caves and leave a free space in front for swimming. The tank should not be excessively lit and the use of ultraviolet lamps is not recommended because the eyes of these fish easily become inflamed. Maintain temperatures of 22° – 28°C (72° – 82°F).
Notes
These fish, very widely distributed in tropical seas, can sometimes be caught at night with a net, though normally with great difficulty. If conditions are correct, they can be kept in captivity for long periods.

NASO LITURATUS — Fam. Acanthuridae

Common name
Japanese tang, lipstick tang.
Dimensions
Up to 40 cm (16 in).
Distribution
Tropical Indo-Pacific region from East Africa to Tuamotu Islands and Hawaii.
Food
In nature: principally herbivorous. In aquarium: seafood and worms, mixed with plant matter and dried fish food likewise containing a high percentage of vegetable matter.
General care
A good aquarium fish. It is as well to keep several individuals of different sizes together; they can also associate with other species. Arrange coral and pieces of rock as shelters. Maintain temperatures of 22° – 28°C (72° – 82°F).
Notes
This species lives on coral reefs, both protected and exposed, usually within the coral branches, down to a depth of 40 m (135 ft). It can be caught with fixed nets.

NASO UNICORNIS — Fam. Acanthuridae

Common name
Unicorn surgeon.
Dimensions
Up to 50 cm (20 in).
Distribution
Tropical Indo-Pacific region from East Africa to Tuamotu Islands and Hawaii.
Food
In nature: herbivorous, usually including seaweed. In aquarium: chopped seafood mixed with fish food rich in vegetable matter, plus seaweed and other algae.
General care
As a rule keep a single specimen together with other species in a large tank, with rocks and corals. Maintain temperatures of 22° – 28°C (72° – 82°F).
Notes
A peculiar feature of this species is the horn on the forehead. In the wild, the fish lives alone or in small groups in coral reef channels, where it can be caught with rod and line, using seaweed as bait.

NEOCLINUS BLANCHARDI — Fam. Clinidae

Common name
Sarcastic fringhead.
Dimensions
Up to 30 cm (12 in).
Distribution
From San Francisco to Baia Magdalena (Mexico).
Food
In nature: carnivorous. In aquarium: chopped seafood, whole *Artemia*.
General care
Lives happily in a large aquarium at temperatures of 13° – 20°C (55° – 68°F). It can live with other species provided these are fairly big.
Notes
This species lives in holes, being most common on steeply inclined walls and reefs. It defends its territory by suddenly opening its huge jaws; it may also attack scuba divers, although it is not capable of causing them serious injury. After locating the fish, it is easy to catch them by means of a fish anaesthetic.

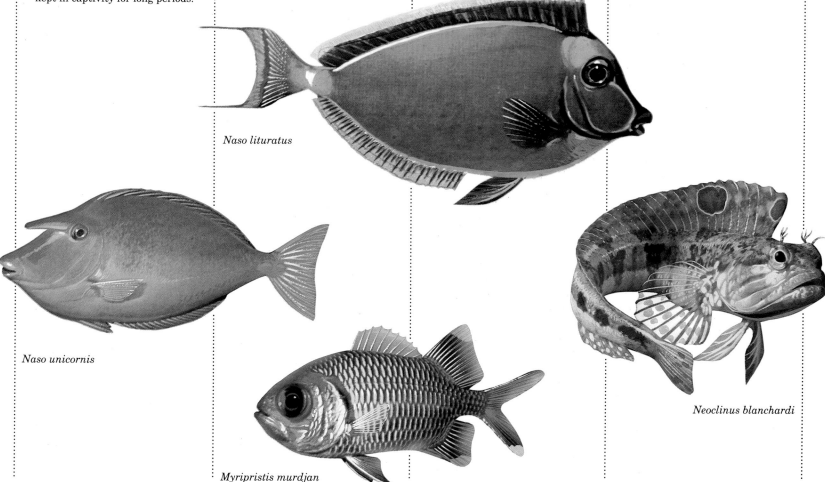

Naso lituratus

Naso unicornis

Myripristis murdjan

Neoclinus blanchardi

OPHIOBLENNIUS STEINDACHNERI
Fam. Blenniidae

Common name
Panama-fanged blenny.
Dimensions
Up to 18 cm (7 in).
Distribution
From Gulf of California to Peru.
Food
In nature: omnivorous; algae and sessile invertebrates. In aquarium: feeds readily on chopped seafood.
General care
Territorial. Keep in an aquarium reserved for this species provided with crevices and rock overhangs, at temperatures of 20° – 23°C (68° – 74°F).
Notes
This blenny has large canine teeth similar to fangs which it uses for protection, as well as sharp teeth that can inflict considerable damage to other fish in the aquarium. In fact, it can even mutilate a bigger fish by ripping off its fins. It is most common on steep rocky shores, in zones of shallow water subject to tidal movement. The young are black and white with red spots and are sometimes mistaken for other species.

OPISTOGNATHUS AURIFRONS
Fam. Opistognathidae

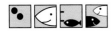

Common name
Yellow-headed jawfish.
Dimensions
Up to 10 cm (4 in).
Distribution
Florida and West Indies.
Food
In nature: carnivorous; small crustaceans and fish. In aquarium: begin with live *Artemia*, then add chopped shrimp, squid and bivalves, krill and other small whole crustaceans as available.
General care
This is a good aquarium fish, to be kept in a tank with one or more other species. Cover the bottom with a deep bed of sand mixed with substrate material of a suitable size for the fish to take in the mouth; add some larger stones. Maintain temperatures of 22° – 27°C (72° – 80°F).
Notes
An attractive and interesting aquarium species which does well in captivity with a suitable substrate. Scuba divers can easily catch it by letting down a small bait into its lair and grabbing the fish as it tries to get rid of the hook.

OPISTOGNATHUS PUNCTATUS
Fam. Opistognathidae

Common name
Fine-spotted jawfish.
Dimensions
Up to 40 cm (16 in).
Distribution
Lake Scammon, Lower California, and from Mexico to Panama.
Food
In nature: small fish and crustaceans. In aquarium: chopped seafood.
General care
As for other species of *Opistognathus*, but the adults need an aquarium of at least 500 liters (110 gal), with more and larger substrate material.
Notes
An excellent aquarium fish for those who have the available space. It is a prodigious burrower, capable of moving a vast amount of material in a single day, especially if dissatisfied with its lair. It can easily be caught at the surface with rod and line while it is searching for a better habitat.

OPISTOGNATHUS RHOMALEUS
Fam. Opistognathidae

Common name
Giant Cortez jawfish.
Dimensions
Up to 60 cm (24 in).
Distribution
Baia Magdalena, Gulf of California and Revillagigedo Islands.
Food
In nature: crustaceans and fish. In aquarium: chopped or whole seafood.
General care
Recommended only to aquarists with large aquaria. These should be of at least 1,000 liters (220 gal), with a bed of 30 cm (12 in) of substrate material for building a lair. Maintain temperatures of 18° – 25°C (65° – 77°F).
Notes
Individuals of this species are often caught by trawlers fishing for prawns, and occasionally by scuba divers. They are interesting aquarium fish, although most aquarists consider them less interesting than other members of the family.

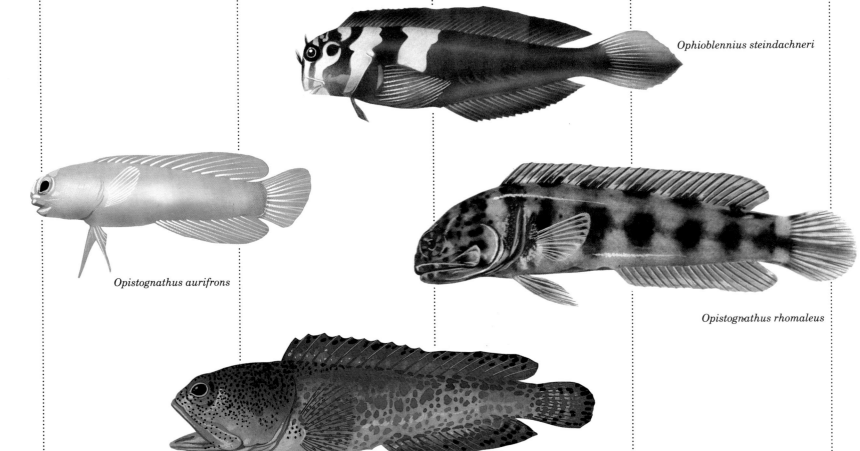

Ophioblennius steindachneri

Opistognathus aurifrons

Opistognathus rhomaleus

Opistognathus punctatus

OSTRACION MELEAGRIS
Fam. Ostraciidae

Common name
White-spotted boxfish.
Dimensions
Up to 15 cm (6 in).
Distribution
Tropical Indo-Pacific region.
Food
In nature: principally benthic invertebrates and algae. In aquarium: fish, chopped shrimp, bivalves, fish and squid.
General care
Suitable only for large aquaria; must be left to acclimate before introducing other fish. Maintain temperatures of 22° – 28°C (72° – 82°F). Partial and regular water changes are advisable.
Notes
Like other fish of this family, *O. meleagris* secretes toxic substances when disturbed. However, if certain precautions are taken, it can be reared successfully.

OSTRACION TUBERCULATUS
Fam. Ostraciidae

Common name
Blue-spotted boxfish.
Dimensions
Up to 45 cm (18 in).
Distribution
Entire Indo-Pacific region.
Food
Omnivorous.
General care
Needs a sandy bottom with rocks and stones for concealment, good lighting and a temperature of 24° – 30°C (75° – 86°F).
Notes
This fish swims with slow, elegant movements. Like many related species, it moves each of its eyes independently, but uses binocular vision when focusing on an object such as a morsel of food.

OXYCIRRHITES TYPUS
Fam. Cirrhitidae

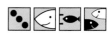

Common name
Long-nosed hawkfish.
Dimensions
Up to 12.5 cm (5 in).
Distribution
Indo-Pacific region and eastern Pacific, including Red Sea, Philippines, Hawaii and Gulf of California.
Food
In nature: little known; probably bottom-swimming plankton, crevice organisms and coelenterate polyps. In aquarium: thrives well on live *Artemia*, mixed with any other acceptable form of fish food.
General care
Although not common, it is an excellent aquarium fish for keeping in a tank of 150 liters (33 gal) or more, at a temperature of around 20°C (68°F), while probably adapting to temperatures of 19° – 27°C (67° – 80°F).
Notes
It is a deep-water species, almost always found at depths of over 30 m (100 ft), on black coral or sea fans. Its bright red markings take on an inconspicuous neutral colour in the wild, as the fish is situated beneath the level of visible red light. Considered rare, it is probably so little known because of the depth of its habitat and its cryptic coloration. In the Gulf of California the species normally lives below the 30-m (100-ft) mark.

OXYMONACANTHUS LONGIROSTRIS
Fam. Ballistidae

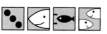

Common name
Orange-spotted emerald filefish, long-nosed filefish.
Dimensions
Up to 10 cm (4 in).
Distribution
Tropical Indo-Pacific region from Red Sea to Samoa.
Food
In nature: principally coral polyps. In aquarium: a varied diet of chopped seafood, small worms and frozen krill, mixed with dried fish food.
General care
Keep several specimens together in a tank reserved only for this species. After they are acclimated and have begun to feed regularly, try putting them in an aquarium together with other species; take care that they have enough to eat, however. Provide shelter with pieces of coral. Recommended temperatures are 22° – 28°C (72° – 82°F).
Notes
This fish normally lives in shallow water among branches of coral and is difficult to catch. Its food poses something of a problem, inasmuch as in the wild the species feeds on coral polyps; so it is a matter of finding a diet that is both nutritious and accepted by the fish.

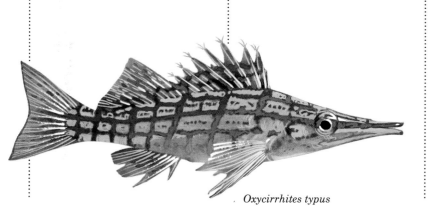

Ostracion tuberculatus

Ostracion meleagris

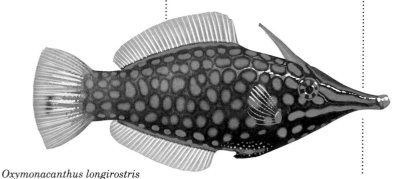

Oxycirrhites typus

Oxymonacanthus longirostris

PARACANTHURUS HEPATUS
Fam. Acanthuridae

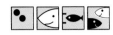

Common name
Regal tang.
Dimensions
Up to 25 cm (10 in).
Distribution
Tropical Indo-Pacific region from East Africa to Line Islands.
Food
In nature: mainly herbivorous. In aquarium: chopped squid and bivalves, fish food containing plenty of vegetable matter, plus the algae that grow in the tank.
General care
It is best to only keep one specimen in a tank together with other species. Prepare an area spacious enough for swimming and arrange various hiding places among the rocks and coral.
Notes
This species is widely distributed but not abundant. In some zones, for example Guam, because of overfishing, it has become rather rare. It is easier to acclimate the young, and they will live peacefully with other fish. It is prone to skin infections. The coloration, brighter in the young, tends to fade with age.

PARAMIA QUINQUELINEATA
Fam. Apogonidae

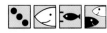

Common name
Five-striped cardinalfish.
Dimensions
Up to 10 cm (4 in).
Distribution
Tropical Indo-Pacific region from East Africa to Line Islands; absent in Hawaii.
Food
In nature: carnivorous, feeding on small fish and crustaceans. In aquarium: a varied diet of live *Artemia*, krill, chopped shrimp and fish food.
General care
Keep several individuals together in a tank of 75 liters (16½ gal) or more, containing just this or other species, too. Prepare caves, crevices and shelters among coral. Maintain temperatures of 22° – 28°C (72° – 82°F).
Notes
The species is found among reefs, in lagoons and around rocks in the open sea at a depth of at least 40 m (135 ft). It lives as a rule in small groups near its places of refuge and is most active at night. The male carries the eggs until they hatch and mouth-broods the young. *P. quinquelineata* is easily caught during the night but must be handled with great care because of its delicacy.

PAREQUES VIOLA
Fam. Sciaenidae

Common name
Rock umbra.
Dimensions
Up to 25 cm (10 in).
Distribution
From Gulf of California to Panama.
Food
In nature: carnivorous and nocturnal; principally crustaceans. In aquarium: shrimp, chopped squid and bivalves; feed the young with live *Artemia* and frozen krill or other whole crustaceans.
General care
Only the young are of interest for home aquaria. They are delicate fish which seek safe hiding places during the day and will live happily only with tranquil companions.
Notes
In summer, groups of young may be observed. The adults almost invariably spend the day hiding in crevices or in the recesses of reefs, emerging only at night to feed.

PERIOPHTHALMUS sp.
Fam. *Gobiidae*

Common name
Mudskipper.
Dimensions
Up to 15 – 30 cm (6 – 12 in).
Distribution
Africa and western Indo-Pacific region.
Food
In nature: carnivorous; small crustaceans and insects. In aquarium: small live crabs and other crustaceans, insects, worms, krill and chopped seafood.
General care
The aquarium must be carefully covered to prevent the fish getting out and should only be partially filled with water. Prepare a substrate of sand or mud with a few stones. To make the environment more authentic, mangroves may also be used. The temperature should be 26° – 30°C (79° – 86°F).
Notes
Because they are so aggressive, mudskippers are usually kept in tanks reserved for their own species. In natural surroundings they spend much of their time out of the water, staying behind to hunt on the beach rather than be carried out with the tide. In fact, they possess an accessory organm which enables them to breathe atmospheric air directly.

Paramia quinquelineata

Paracanthurus hepatus

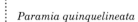

Pareques viola

PHOTOBLEPHARON PALPE-BRATUS Fam. Anomalopidae

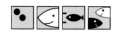

Common name
Lanternfish, flashlight.
Dimensions
Up to 9 cm (3½ in).
Distribution
Red Sea and Indian Ocean.
Food
In nature: carnivorous; pelagic crustaceans and small fish. In aquarium: live *Artemia* and other live crustaceans, frozen krill.
General care
As for *Anomalops*.
Notes
It may easily be distinguished from other species in that it has a single dorsal fin. It adapts quite well to captivity, as does *Kryptophanaron alfredi*. A specimen of *K. harveyi* has been found in the eastern Pacific but this has never been kept in an aquarium. The luminous organ depends on the presence of luminescent bacteria which produce cold light by means of a mechanism akin to that of glow-worms. The state of the fish's health can be assessed by its capacity to emit light, which, in fact, tends to diminish as soon as the fish is weakened by illness or unsuitable environmental conditions.

PLATAX ORBICULARIS Fam. Ephippididae

Common name
Roundfin batfish.
Dimensions
Up to 50 cm (20 in).
Distribution
Tropical Indo-Pacific region; not found in eastern Pacific and Hawaii.
Food
In nature: omnivorous, including refuse. In aquarium: accepts ordinary dried fish food, but it is advisable to mix this with chopped molluscs.
General care
It needs room to grow. In captivity, especially at the start, it may be timid, but it soon becomes tame. It must not be left with species that will bite its fins. Maintain temperatures of 22° – 28°C (72° – 82°F).
Notes
P. orbicularis is a coastal species which is frequently seen in bays and harbours. The young assume the cryptic guise of dead leaves, lying or floating on one side in order to avoid the notice of predators. As a rule the adults are gregarious, forming shoals of up to 30 individuals. Raising them in aquaria presents no great difficulties, particularly if young and healthy specimens, which have been suitably quarantined, are chosen. This condition, incidentally, applies to all seawater species. Obviously, to keep this species in captivity, account must be taken of their potential size and the fact that as a result of the large quantities of waste matter they produce, small aquaria may prove unsuitable.

PLECTORHYNCHUS CHAETO-DONOIDES Fam. Haemulidae

Common name
Sweetlips, polkadot fish.
Dimensions
Up to 45 cm (18 in).
Distribution
Tropical Indo-Pacific region from East Indies to Tahiti.
Food
In nature: omnivorous. In aquarium: fish, bivalves and chopped squid, mixed, if desired, with vegetable matter.
General care
The young live best in tanks with other species, provided there are plenty of hiding places. Maintain temperatures of 22° – 28°C (72° – 82°F).
Notes
Only the young are of real interest to aquarists because the adults are more uniform in colour, with small spots, and less handsome. The young are characterized by large white blotches which persist until the fish reaches a length of 13 cm (5¼ in). In the opinion of many aquarists, this species is not easy to raise. Above all, a great deal of patience is required to find the most suitable type of diet, which should be as varied as possible.

PLOTOSUS ANGUILLARIS Fam. Plotosidae

Common name
Saltwater catfish.
Dimensions
Up to 70 cm (28 in).
Distribution
Indo-Pacific region from East Africa to Polynesia.
Food
In nature: bottom-dwelling organisms and organic material. In aquarium: chopped fish and meat, worms, dried fish food.
General care
Only the young are suited to home aquaria. Keep several individuals together in an aquarium with a sandy substrate; they may live with other varieties. Maintain temperatures of 20° – 30°C (68° – 86°F).
Notes
This is an estuarine species, very common in coastal waters of the Indo-Pacific zone.

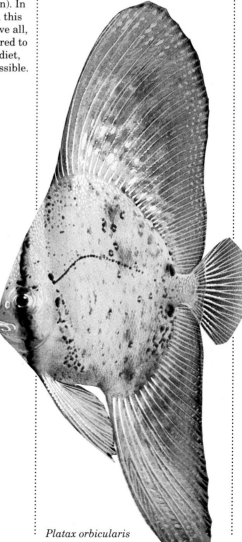

Platax orbicularis

Plotosus anguillaris

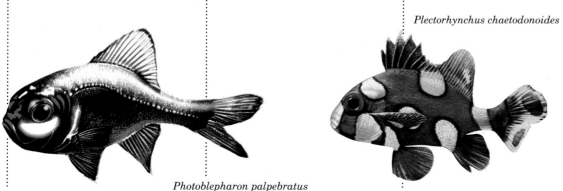

Plectorhynchus chaetodonoides

Photoblepharon palpebratus

POLYDACTYLUS OLIGODON
Fam. Polynemidae

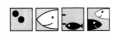

Common name
Smallscale threadfin.
Dimensions
Up to 22 cm (8½ in).
Distribution
From Florida to Brazil.
Food
In nature: shrimp, crabs and interstitial fauna. In aquarium: chopped seafood.
General care
Use a large aquarium with a sandy bed for the young. For adults use an aquarium of 2,000 liters (440 gal) or more.
Notes
This species normally lives in sandy areas and often allows itself to be carried by the breakers on to the beach. Here they seek food thrown up by wave action, which they capture expertly, thanks to the underside of their pectoral fins, which, typical of the family, is furnished with 3–7 elongated and separated rays that serve as sensory organs.

POMACANTHUS ARCUATUS
Fam. Pomacanthidae

Common name
Grey angelfish.
Dimensions
Up to 50 cm (20 in).
Distribution
Tropical western Atlantic from Brazil to Florida, sometimes found as well off New England.
Food
In nature: omnivorous, with a preference for sponges. In aquarium: give a varied diet comprising squid and fish, mixed with flaked fish food.
General care
The adults require large aquaria of at least 2,000 liters (440 gal). Maintain temperatures of 20° – 28°C (68° – 82°F).
Notes
Individuals of this species are common in and around coral reefs down to a depth of 30 m (100 ft) and normally living alone or in pairs and being readily approachable. The young are very similar to those of their close relative *Pomacanthus paru*, and both species may live for several years in captivity.

POMACANTHUS IMPERATOR
Fam. Pomacanthidae

Common name
Emperor angelfish.
Dimensions
Up to 31 cm (12¼ in).
Distribution
Indo-Pacific region from Africa to Pitcairn Island; rare in Hawaii.
Food
In nature: sponges, tunicates and other invertebrates. In aquarium: quickly becomes accustomed to a varied diet.
General care
It requires a big tank of 500 liters (110 gal) or more, at a temperature of 24° – 28°C (75° – 82°F).
Notes
This is an extremely popular aquarium fish because of its superb coloration and hardiness. It normally lives alone or in pairs outside coral reefs, down to a depth of 24 m (80 ft).
If not kept in pairs, it displays aggressiveness towards members of its own species and other large species of Pomacanthidae. If kept with other angelfish, individuals should differ in length by 2 – 3 cm (¾ – 1¼ in) so as to reduce likelihood of fighting.

POMACANTHUS MACULOSUS
Fam. Pomacanthidae

Common name
Moon angelfish.
Dimensions
Up to 30 cm (12 in).
Distribution
Indian Ocean, including Red Sea and coasts of East Africa.
Food
In nature: principally sponges and tunicates. In aquarium: provide a varied diet.
General care
It is best to use a large aquarium of 500 liters (110 gal) or more; maintain temperatures of 24° – 28°C (75° – 82°F).
Notes
This species is easy to keep. Provide the tank with plenty of hiding places.

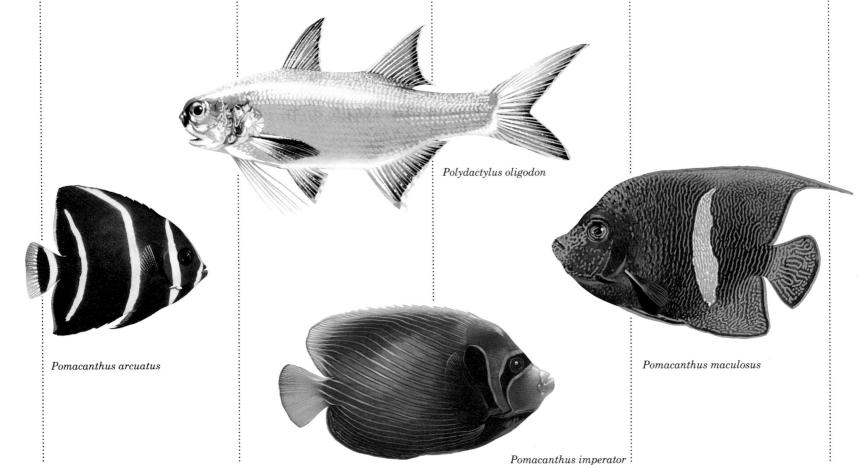

Polydactylus oligodon

Pomacanthus arcuatus

Pomacanthus imperator

Pomacanthus maculosus

POMACANTHUS PARU
Fam. Pomacanthidae

Common name
French angelfish.
Dimensions
Up to 38 cm (15 in).
Distribution
Indo-Pacific region from Africa to Pitcairn Island; rare in Hawaii.
Food
In nature: sponges, tunicates and other invertebrates. In aquarium: normally adapts quickly and successfully to a very varied diet.
General care
As for *P. arcuatus*. It is useful to prepare a busy bottom with many hiding places where this territorial species can find refuge.
Notes
This species greatly resembles *P. arcuatus* but differs in the coloration of the caudal fin, which in the latter terminates in a broad, almost transparent, band, while the caudal peduncle is adorned with a yellow stripe. Young individuals behave like cleaner fish, ridding other species of parasites.

POMACANTHUS ZONIPECTUS
Fam. Pomacanthidae

Common name
Cortez's angelfish.
Dimensions
Up to 45 cm (18 in).
Distribution
Eastern Pacific from Gulf of California to Peru.
Food
In nature: feeds on small invertebrates. In aquarium: eats a large variety of seafood, at great speed.
General care
Adults need a very large aquarium of 2,000 liters (440 gal) or more, maintained at 20° – 24°C (68° – 75°F). The young can be kept in a tank of 500 liters (110 gal) or more.
Notes
The illustration shows an adult specimen. The young are dark blue with curving blue and yellow stripes along the flanks.

PRIONURUS PUNCTATUS
Fam. Acanthuridae

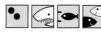

Common name
Yellow-tailed surgeonfish.
Dimensions
Up to 60 cm (24 in).
Distribution
From Gulf of California to El Salvador, including the Revillagigedo Islands (Mexico).
Food
In nature: herbivorous. In aquarium: accepts a wide variety of food such as fish, squid, bivalves, etc. Mix, if possible, with vegetable matter.
General care
An extremely sturdy species. Only the young are suitable for a normal home aquarium, although in captivity they do not grow much. They are interesting to keep in association with other species.
Notes
Adults and juveniles form shoals at depths of 3 – 12 m (10 – 40 ft) and occasionally down to 30 m (100 ft) or more. They can be caught by means of a fixed net or by scuba divers with a hand net, at night. The young are yellow and very similar to *Zebrasome flavescens*.

PSEUDOCHROMIS FLAVIVERTEX
Fam. Pseudochromidae

Common name
Orange-striped dottyback.
Dimensions
Up to 10 cm (4 in).
Distribution
Tropical Indo-Pacific region.
Food
In nature: plankton. In aquarium: live and newly born *Artemia*, mixed with dried fish food.
General care
This fish is happy to live with other species. The tank must have many small rock and coral holes to act as hiding places. Maintain temperatures of 22° – 28°C (72° – 82°F).
Notes
This is a tranquil, active species, well adapted to aquarium life, although *P. porphyreus* is an easier species to raise. Keep several individuals together because they are gregarious. They do not like too much light and should not be kept with invertebrates, which they will end up eating.

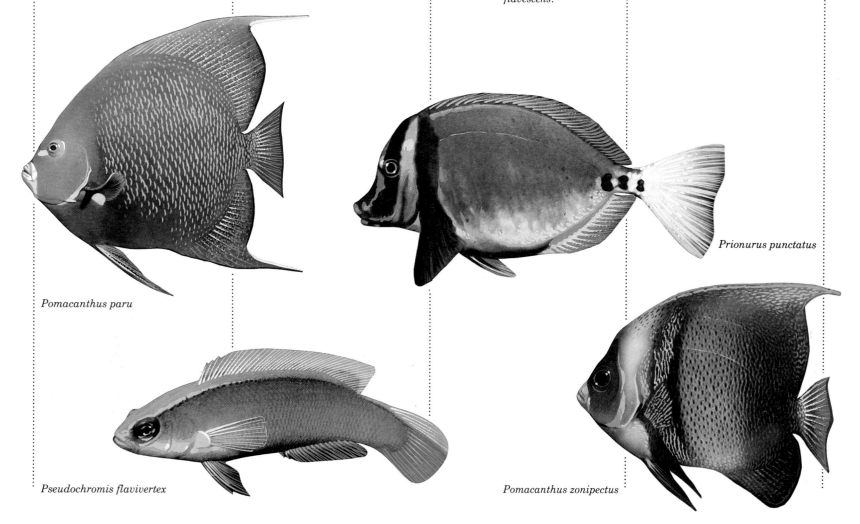

Pomacanthus paru

Prionurus punctatus

Pseudochromis flavivertex

Pomacanthus zonipectus

PSEUDOCHROMIS PORPHYREUS
Fam. Pseudochromidae

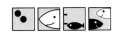

Common name
Purple dottyback.
Dimensions
Up to 6 cm (2½ in).
Distribution
Tropical Indo-Pacific region.
Food
In nature: bottom-dwelling plankton. In aquarium: live *Artemia*, small worms, krill, fish food.
General care
Adapts to living in tanks with other species, but needs numerous hiding places among rocks and coral. Keep 2–3 specimens in a 75-liter (16½-gal) aquarium at temperatures of 22° – 28°C (72° – 82°F).
Notes
This species is usually to be found along the steep walls of coral reefs at a depth of 15 m (50 ft) or more. As a rule, the fish stay several feet apart from one another. They are difficult to catch, even with a fish anaesthetic. Make sure that specimens from the Philippines have not been caught with cyanide because they do not normally survive long.

PTEROIS VOLITANS
Fam. Scorpaenidae

Common name
Scorpion fish, lionfish.
Dimensions
Up to 30 cm (12 in).
Distribution
Tropical Indo-Pacific region, excluding Hawaii.
Food
In nature: carnivorous; fish and crustaceans. In aquarium: chopped fish, squid and shrimp; try also to include live fish in the diet.
General care
This fish is easy to keep, but grows rapidly, so you will need a tank of at least 200 liters (44 gal) to contain several individuals. It will live with other species provided they are not so small as to provide a meal. Maintain temperatures of 22° – 30°C (72° – 86°F).
Notes
This species lives in lagoons, channels and partially protected reefs, at a depth of 35 m (115 ft) or more. As a rule, they can be seen under the reef, preying on small fish and crustaceans. Its beautiful wavy fins are surmounted by sharp, poisonous spines, which are used as a means of defense and may cause painful though not fatal injuries.

PYGOPLITES DIACANTHUS
Fam. Pomacanthidae

Common name
Regal angelfish.
Dimensions
Up to 25 cm (10 in).
Distribution
Indo-Pacific region from East Africa and Red Sea to Marquesas.
Food
In nature: mainly sponges and tunicates. In aquarium: a varied diet is essential; provide chopped shrimp and squid, live *Artemia*, flaked fish food and vegetable matter. Fish food in pellet form is also useful.
General care
Because this is a difficult species to keep, it is recommended only to expert aquarists. Adults need a large aquarium of 500 liters (110 gal) or more, at a temperature of 24° – 28°C (75° – 82°F). Create caves and crevices in which the fish can hide.
Notes
This species is common along the Great Barrier Reef but is normally fished for export in the Philippines and Sri Lanka. Specimens from the latter region are recommended for the aquarium. As a rule, it lives in coral-rich zones, in shallow water, not deeper than 25 m (80 ft).

RHINECANTHUS ACULEATUS
Fam. Balistidae

Common name
Picasso trigger.
Dimensions
Up to 30 cm (12 in).
Distribution
Tropical Indo-Pacific region from East Africa to Polynesia, including Hawaii.
Food
In nature: carnivorous; it feeds on crustaceans and other invertebrates. In aquarium: chopped fish, squid and bivalves, plus other fish food, if desired.
General care
As a rule keep a single individual with other species, although young specimens can live together in a community aquarium. Create hiding places and maintain temperatures of 22° – 28°C (72° – 82°F).
Notes
This is a very popular aquarium fish. It is widely found in shallow protected zones and can be caught in crevices.

Pygoplites diacanthus

Pterois volitans

Rhinecanthus aculeatus

Pseudochromis porphyreus

RHINECANTHUS RECTANGULUS
Fam. Balistidae

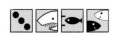

Common name
Rectangular trigger.
Dimensions
Up to 23 cm (9 in).
Distribution
Tropical Indo-Pacific region from East Africa to Polynesia, including Hawaii.
Food
In nature: omnivorous; its diet comprises crustaceans, sea urchins and algae. In aquarium: chopped seafood, mixed, if desired, with vegetable matter.
General care
It is possible to keep one individual together with other species, but it must be removed immediately if problems arise. Recommended temperatures are 22° – 28°C (72° – 82°F).
Notes
The adults live outside the coral reef, in zones subject to currents. The young are found in reef pools. The fish can be caught by collecting the rocks where they have their lair.

RHINOMURAENA AMBOINENSIS
Fam. Muraenidae

Common name
Blue ribbon eel.
Dimensions
Up to 125 cm (50 in).
Distribution
Tropical western Pacific.
Food
In nature: carnivorous; crustaceans and fish are included in its diet. In aquarium: chopped seafood mixed with small live fish.
General care
Prepare holes and crevices as hiding places, against a dark background. Keep the light dim and turn it off at night. The fish can live quite well in an aquarium with other species but not with those that are aggressive, such as territorial Pomacentridae, or with small fish that may constitute a meal. The aquarium, of 125 liters (27½ gal) or more, should be kept at a temperature of 24° – 29°C (75° – 84°F). Although the species is familiar to aquarists, it is more difficult to keep than others. Make sure the holes prepared are big enough for complete concealment. The diet should be varied, including small fish and, if possible, small live crustaceans.
Notes
R. amboinensis is normally found in shallow lagoons among coral. When buying a specimen, make sure it has not been caught with cyanide.

SCARUS GHOBBAN
Fam. Scaridae

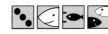

Common name
Parrotfish.
Dimensions
Up to 80 cm (32 in).
Distribution
Tropical Indo-Pacific region.
Food
In nature: algae, corals and other organisms living on them. In aquarium: a varied diet, including vegetable matter. Some aquarists have had success by fixing the vegetation to supports similar to branches of coral.
General care
No matter which species of parrotfish you choose, always go for young subjects. Use a large aquarium with a high layer of sand and a few stones.
Notes
Common in shallow tropical waters, they are active by day and hide in caves and crevices at night; some species secrete a cocoon of mucus which can eliminate their odour, preventing them being located. Polychromatism is common among the Scaridae; males and females, like the young, may have completely different colours. Generally, in *Scarus ghobban* the males differ in colour from the females, while the young of both sexes have the same coloration as the adult females.

SCORPAENODES XYRIS
Fam. Scorpaenidae

Common name
Rainbow scorpion fish.
Dimensions
Up to 18 cm (7 in).
Distribution
From southern California to Peru.
Food
In nature: carnivorous; feeds on small fish, crustaceans, etc. In aquarium: fish, chopped squid and shrimp.
General care
A very easy fish to keep, particularly in tanks with other species, provided they are not too small and thus tempting to eat. Recommended temperatures are 20° – 27°C (68° – 80°F).
Notes
This small fish is highly camouflaged, relatively inactive, but very sturdy. It is therefore a useful addition to the aquarium with other species. It is found along rocky coasts with numerous crevices. Once located, it is easily caught with a net and probe. For reasons unknown, the largest specimens are present in the extreme northern part of distribution zones.

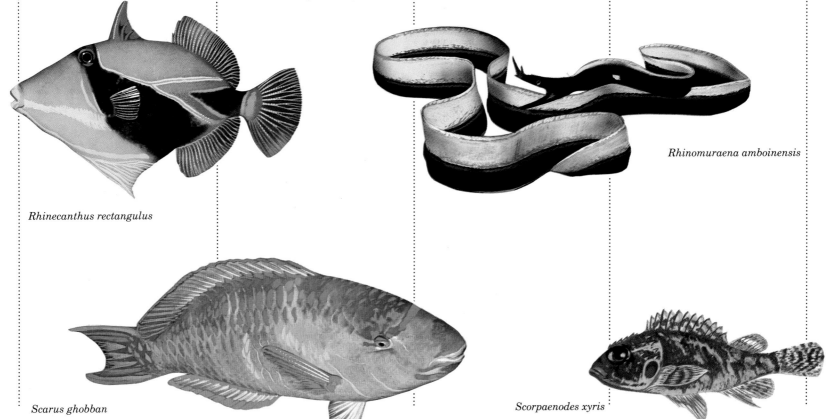

Rhinecanthus rectangulus

Rhinomuraena amboinensis

Scarus ghobban

Scorpaenodes xyris

TROPICAL SEA WATERS • 105

SERRANUS FASCIATUS
Fam. Serranidae

Common name
Banded comber.
Dimensions
Up to 11 cm (4½ in).
Distribution
From Gulf of California to Peru and Galapagos Islands.
Food
In nature: carnivorous, appearing to feed on small fish and crustaceans. In aquarium: chopped seafood.
General care
A hardy species, relatively inactive and quiet. It adapts to life with other species but should not be kept with other small fish. Maintain temperatures of 20° – 25°C (68° – 77°F).
Notes
It is a shallow water species, found along rocky shores and in zones with thick vegetation. It is quite easy to catch and, given that it can be fished at less than 6 m (20 ft), does not need to be degassed. It is inoffensive by nature and therefore should not be kept with aggressive Pomacentridae. Others of the genus include the small species *Serranus tigrinis* (tropical Atlantic) and *S. annularis*, brightly coloured and easy to raise in the aquarium.

SERRANUS TABACARIUS
Fam. Serranidae

Common name
Tobacco fish.
Dimensions
Up to 15 cm (6 in).
Distribution
From Bermuda to Lesser Antilles.
Food
In nature: carnivorous. In aquarium: pieces of meat, squid, shrimp and bivalves; small live fish.
General care
It will adapt to living in tanks with other fish of equal or bigger size. Make hiding places among rocks. Maintain temperatures of 21° – 24°C (70° – 75°F).
Notes
This species varies greatly in coloration but may also possess little colour. One of the smallest of the Serranidae available to collectors, it fares well in captivity. Solitary by nature, it is usually caught close to the coral reef, swimming just above the seabed. It has been caught at depths of up to 40 m (135 ft).

SPHAERAMIA NEMATOPTERUS
Fam. Apogonidae

Common name
Red-spotted cardinal.
Dimensions
Up to 10 cm (4 in).
Distribution
Tropical Indo-Pacific region from East Africa to Fiji.
Food
In nature: small fish and large plankton. In aquarium: shrimp, squid and chopped fish, krill, live *Artemia*, fish food.
General care
A quiet species. Keep several individuals together in a tank provided with hiding places among rocks and coral. It is delicate and probably thrives best in an aquarium on its own. Maintain temperatures of 22° – 28°C (72° – 82°F).
Notes
This fish must be handled very carefully when being caught and transported, and kept only with tranquil species. It is commonly found in protected lagoons between coral reefs and mangrove swamps. The young are bright yellow and have red spots that disappear with growth. The females have an elongated dorsal fin and more distinctive markings. The young are mouth-brooded and transported by the male.

STEGASTES FLAVILATUS
Fam. Pomacentridae

Common name
Beaubrummel.
Dimensions
Up to 10 cm (4 in).
Distribution
From Gulf of California to Ecuador.
Food
As for *S. rectifraenum*.
General care
As for *S. rectifraenum*. The young are aggressive but can be kept with large active fish. Prepare suitable shelters and an open space for swimming.
Notes
Adults of this species are distinguished from *S. rectifraenum* by the yellowish margins to the dorsal, anal and caudal fins. The young are easily recognized by the lower part of the body, which is bright yellow. Environment and behaviour are similar to those of *S. rectifraenum*, and similarly this fish is markedly territorial, not hesitating to attack even larger fish whenever they come too close for comfort.

Serranus fasciatus

Stegastes flavilatus

Sphaeramia nematopterus

Serranus tabacarius

STEGASTES RECTIFRAENUM
Fam. Pomacentridae

Common name
Cortez damselfish.
Dimensions
Up to 13 cm (5¼ in).
Distribution
Gulf of California and coasts of Lower California to Baia Magdalena (Mexico).
Food
In nature: mainly herbivorous. In aquarium: adapts to various diets.
General care
The adults, inconspicuous and aggressive, are not often kept in the aquarium. It is essential to prepare many hiding places (caves, holes and crevices) in a large 400-liter (88-gal) tank. They are sturdy and tolerate water temperatures of 20° – 28°C (68° – 82°F). It is best to only keep one individual or a pair.
Notes
The adults are markedly territorial, but strangely, it has been verified in the course of experiments that such behaviour ceases when the room temperature is maintained at 12°C (53°F). During courtship and egg laying, the female takes on a light reddish brown colour on the head and dorsal fin, while the lips turn yellow. Because the genus *Stegastes* is a fairly recent attribution, some books still list this species under its former name of *Eupomacentros rectifraenum*.

SUFFLAMEN VERRES
Fam. Balistidae

Common name
Orangemice triggerfish.
Dimensions
Up to 38 cm (15½ in).
Distribution
From Lower California to Ecuador.
Food
In nature: carnivorous, especially crustaceans, sea urchins, molluscs and worms. In aquarium: chopped seafood; thrives well on a diet of fish and squid.
General care
Lives happily in an aquarium with other species, at temperatures of 20° – 24°C (68° – 75°F).
Notes
It is a sturdy species, adapting easily to captivity, and very long-lived. Young specimens are preferable for home aquaria but are not abundant in the wild. Active by day, these fish are normally solitary and shelter at night in crevices. They can be caught both during the day and night by releasing the dorsal fin and removing the fish from their holes.

SYNANCEIA VERRUCOSA
Fam. Synanceiidae

Common name
Stonefish.
Dimensions
Up to 33 cm (13 in).
Distribution
Tropical Indo-Pacific region from East Africa to Tahiti, including Red Sea; not found in Queensland or Hawaii.
Food
In nature: carnivorous; small fish and crustaceans. In aquarium: live fish, fresh or frozen fish and shrimp.
General care
Use a firmly covered, escape-proof aquarium. This fish adapts well to life with other species, provided they are not so small that they may be eaten. It is sedentary, requiring very little space. A 25-cm (10-in) specimen can be kept in a 75-liter (16½-gal) tank. Maintain temperatures of 22° – 28°C (72° – 82°F).
Notes
Many other species of Synanceiidae have been kept in aquaria, such as *Synanceia horrida*, *Inimicus didactylus*, etc. All are dangerous to handle.

SYNCHIROPUS PICTURATUS
Fam. Callionymidae

Common name
Psychedelic fish.
Dimensions
Up to 10 cm (4 in).
Distribution
Tropical western Pacific.
Food
In nature: carnivorous. In aquarium: small worms, *Tubifex*, crustaceans, live *Artemia*, perhaps dried fish food.
General care
Can live with other tranquil species, but aquarists often prefer to keep one individual or a pair as single species. Cover the bottom with sand and make crevices for hiding. Maintain temperatures of 22° – 28°C (72° – 82°F).
Notes
This species is usually found in calm, shallow waters. At Palau it is commonly found in inland lagoons at the base of cliffs where it can be collected with a net or by using fish anaesthetic.

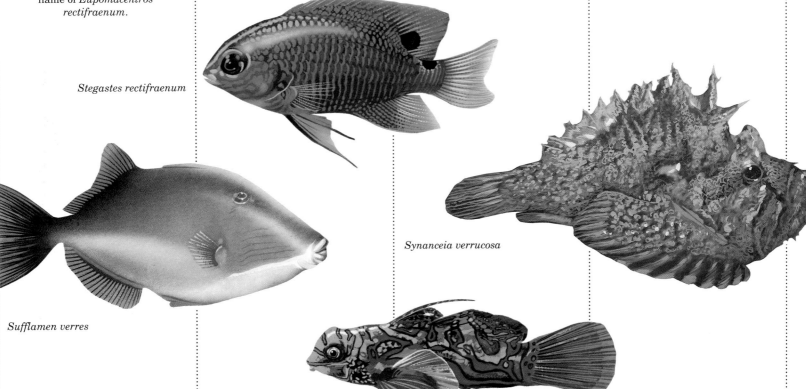

Stegastes rectifraenum

Sufflamen verres

Synanceia verrucosa

Synchiropus picturatus

TAENIOCONGER DIGUETI
Fam. Congridae

Common name
Cortez's sand conger.
Dimensions
Up to 76 cm (30 in).
Distribution
Gulf of California, Mexico.
Food
In nature: invertebrates of various genera (not only benthic) and fish. In aquarium: small specimens accept adult *Artemia*, mixed with other plankton and worms; once acclimated, they will also take dead organisms of a size commensurate to themselves.
General care
Use a large, high aquarium of at least 600 liters (132 gal), equipped with an undergravel filter. Cover the bottom with loose material such as coral sand, to a thickness of 15 cm (6 in) or more.
Notes
This is not a suitable fish for beginners. It should be kept in a tank reserved for this species, which should not be too brightly illuminated.

THERAPON JARBUA
Fam. Theraponidae

Common name
Tigerperch.
Dimensions
Up to 25 cm (10 in).
Distribution
Tropical Indo-Pacific region from East Africa to Melanesia and Taiwan.
Food
In nature: carnivorous. In aquarium: squid, bivalves and shrimp; also small live fish and fish food, if desired.
General care
It is easy to keep, shoaling contentedly in tanks reserved for its own kind but adapting, too, to the presence of other species. The temperature should be 22° – 28°C (72° – 82°F); it also adapts to brackish water.
Notes
Although this fish reacts well to captivity, it is not too popular with aquarists because of its insignificant appearance. It is abundant in bays and estuaries, feeding on small fish and other soft-bodied organisms. It can be caught quite easily with a net or rod and line. It has been bred successfully in the aquarium.

THALASSOMA LUCASANUM
Fam. Labridae

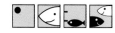

Common name
Cortez's rainbow wrasse.
Dimensions
Up to 10 cm (4 in).
Distribution
Gulf of California to Panama, Galapagos Islands.
Food
In nature: small benthic invertebrates and bottom-swimming plankton; also acts as a cleaner fish. In aquarium: chopped shrimp and other types of seafood.
General care
This is an excellent aquarium fish which adapts to life with other species. Young individuals can be kept in a 75-liter (16½-gal) tank at temperatures of 20° – 28°C (68° – 82°F), with a well-aerated sandy bed and rocks and coral for shelter.
Notes
Shoals of *Thalassoma* are often seen around shallow coral reefs. To facilitate their capture with nets and traps, pieces of sea urchin can be used as bait. The stage in which the fish has a blue head is secondary (caused by sexual inversion), while the normal (rainbow) colour is present in young and adult males and females.

TRIAKIS SEMIFASCIATA
Fam. Carcharhinidae

Common name
Leopard shark.
Dimensions
Up to 2 m (6½ ft).
Distribution
Oregon to Gulf of California and Mexico.
Food
In nature: carnivorous; benthic invertebrates, crustaceans, small benthic fish. In aquarium: soon learns to feed on fresh or frozen fish and squid.
General care
Only young specimens are suitable for home aquaria. Use a tank at least three times the length of the animal. Maintain temperatures of 13° – 24°C (55° – 75°F).
Notes
In the wild the species lives close to sandy and muddy seabeds, to a depth of 100 m (330 ft). Related to this species are *Mustelus californicus*, *M. mustelus* and *M. asterias*, the last two also found in the Mediterranean, but none of which are regular subjects of aquaria.

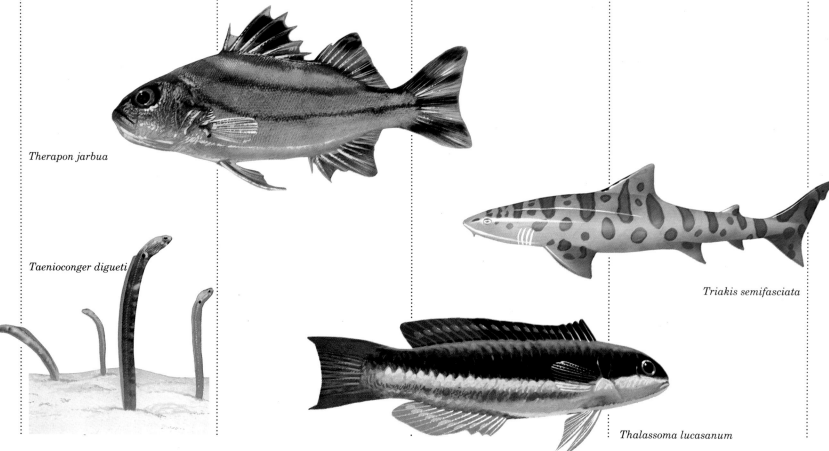

Therapon jarbua

Taenioconger digueti

Triakis semifasciata

Thalassoma lucasanum

XANTHICHTHYS MENTO
Fam. Balistidae

Common name
Pink-tailed triggerfish.
Dimensions
Up to 30 cm (12 in).
Distribution
Tropical Pacific.
Food
In nature: omnivorous. In aquarium: feeds on almost anything that is offered.
General care
A sturdy, active species, suitable for spacious aquaria containing large fish. Maintain temperatures of 20° – 23°C (68° – 74°F).
Notes
This fish is very common along the outer walls of coral reefs and close to marine escarpments, where it is often pelagic. Although it generally preys on shelled invertebrates, it will feed on virtually any organic material, including refuse thrown overboard from ships. It can be caught easily with line and bait. A close relative, *X. ringens*, lives in the Atlantic and is an excellent aquarium fish.

ZANCLUS CANESCENS
Fam. Acanthuridae

Common name
Moorish idol.
Dimensions
Up to 23 cm (9 in).
Distribution
Tropical Indo-Pacific region from East Africa to Panama, including Hawaii and Gulf of California.
Food
In nature: omnivorous (sponges, coral algae and crevice-dwelling invertebrates). In aquarium: sometimes difficult to feed. Try chopped shrimp, squid and bivalves, together with live *Artemia*, krill, small worms and vegetable matter. Usually accepts broccoli.
General care
It is possible to keep several individuals together in an aquarium with different species provided these are large enough not to be eaten. They thrive best in large tanks with plenty of space for swimming and hiding places with several entrances. Temperatures should be 20° – 28°C (72° – 82°F).
Notes
This species is well-known but can be a problem to keep in captivity. It can easily be caught at night by scuba divers with hand nets.

ZEBRASOMA FLAVESCENS
Fam. Acanthuridae

Common name
Yellow tang.
Dimensions
Up to 20 cm (8 in).
Distribution
Tropical zones of western and central Pacific from southern Japan to Hawaii.
Food
In nature: filamentous algae. In aquarium: chopped seafood and *Artemia* mixed with vegetable matter.
General care
They are territorial and tend to fight with those of their species. Try to keep one or two together with other varieties, at temperatures of 21° – 24°C (70° – 75°F). They thrive best in large tanks illuminated by sunlight.
Notes
This species lives at depths of up to 30 m (100 ft). Shallow water specimens feed to a large extent on green algae while those inhabiting deep water may be omnivorous.

ZEBRASOMA VELIFERUM
Fam. Acanthuridae

Common name
Striped sailfin tang.
Dimensions
Up to 40 cm (16 in).
Distribution
Indo-Pacific region.
Food
Much vegetable matter (in the wild it grazes on seagrass); not easy to domesticate.
General care
Needs a great deal of free space for swimming and hiding, on a high layer of sand. The water temperature should be 23° – 26°C (74° – 79°F); it can be helpful to provide an ozone reactor and ultraviolet sterilizer.

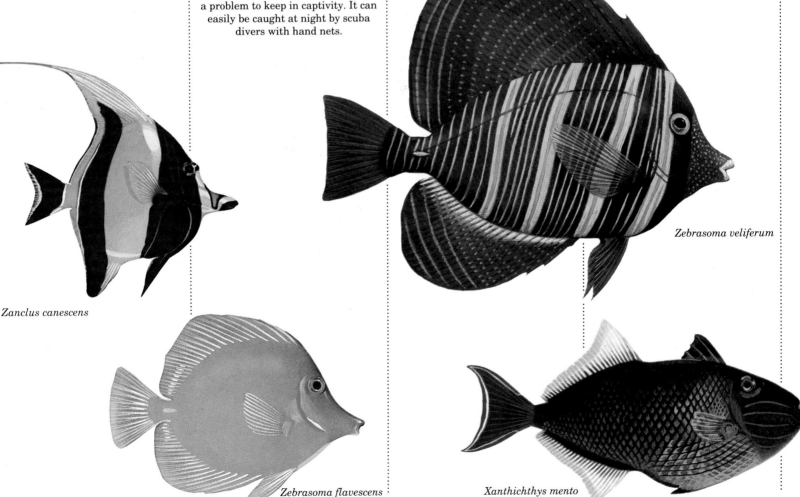

Zanclus canescens

Zebrasoma veliferum

Zebrasoma flavescens

Xanthichthys mento

TEMPERATE SEA WATERS

Temperate sea waters

The rocky shores of the Mediterranean (below) and submerged reefs (opposite, below) constitute habitats that, with their wealth of living forms, are full of interest for aquarists.

Observed from satellites, our earth justly deserves the description "blue planet" – the name given it by early astronauts. The oceans of the world cover, in fact, approximately 360 million km² (139 million sq. miles), representing about 70 per cent of the earth's surface. They occupy a volume of some 1,370 million km³ (324 million cu. miles), with a mean depth of 3,800 m (12,500 ft), including points such as the Guam Trench, where the depth is 11,500 m (37,000 ft).

For aquarists, the interesting areas of "blue" consist of tropical seas and only a small part is represented by temperate seas. The latter tend to arouse even less interest and enthusiasm among aquarists than the more confined freshwater habitats. The reasons are probably threefold: there are comparatively few fish species commercially available, they are not particularly well known, and they are not easy to keep successfully in the aquarium. Apart from the fact that they exhibit marked differences in light penetration, salinity and concentration of nutrients, temperate sea waters may be distinguished from their tropical counterparts mainly with regard to mean temperature, which fluctuates around 18° – 20°C (65° – 68°F) or below. Of all temperate seas, certainly the best known, both by reasons of the important events that have occurred and the civilizations that have flourished along its shores, is the Mediterranean. Bridging, rather than separating, the continents of Europe, Africa and Asia, the geological history of the Mediterranean is complex, for it was once part of the Tethys Sea, communicating with the Pacific Ocean. Eventually it was transformed into a gigantic lake, isolated at first, then connected to the Atlantic Ocean and, more recently as a result of human engineering skill, to the Red Sea. These happenings have had their effect on the fauna of the region, so that not only are there endemic species, but also species with tropical and subtropical features, as well as others with Atlantic affinities. And this does not take into account the so-called Lessepsian migrations (named after the Frenchman Ferdinand Lesseps, builder of the Suez Canal) whereby dozens of Red Sea species have travelled through the canal to colonize the Mediterranean. Approximately 3,800 km (2,360 miles) long and 1,800 km (1,120 miles) across at its widest point, the Mediterranean owes its classification as a "warm" temperate sea to the existence of the sill of Gibraltar, situated about 300 m (1,000 ft) beneath the surface. This prevents the deeper and colder waters of the Atlantic mixing with those of the Mediterranean, the seabed temperature of which remains constantly at around 13°C (55°F). Only the surface waters of the Atlantic enter the Mediterranean, through the Strait of Gibraltar, so that the deep exit currents involve exclusively Mediterranean water. The presence of the Italian peninsula, moreover, which divides the Mediterranean into two basins, west and east, further complicates the hydrography of this enclosed sea.

The surface temperature, both inshore and farther out, never varies by more than 7° – 8°C (9° – 12°F) over its entire extent. Temperature, as everywhere else, is the environmental parameter that has the most profound influence upon all forms of aquatic life; and this factor has to be considered when setting up an aquarium. In the case of the majority of temperate water organisms kept in captivity, this means that the water has to be cooled rather than heated. This necessitates the installation of a cooling system, something that is too expensive for most amateur aquarists. The only practical alternative for those who are keen on collecting temperate marine fish is to place the tank in a cool place which receives little if any heating, such as a cellar or a basement. This, of course, will limit one's enjoyment of the aquarium and its occupants.

The need to maintain a low water temperature is thus the main reason why comparatively few fish collectors choose to keep Mediterranean and other temperate marine species. Even so, it is worth describing some typical environments, at the same time offering a few practical suggestions for setting up such aquaria in the home.

The surface strip that divides the land from the sea is covered and exposed by the tides at regular intervals, and it contains a vast number of very different groups of organisms. The plants and animals of the tidal zone have evolved and adapted to this particular habitat, often developing peculiar and distinctive structures and behaviour patterns. Barnacles and limpets, for example, manage to survive exposure by retaining in their shell a stock of water sufficient for the hours that they have to spend on dry land. Other organisms, regulated by an invisible inner clock, anticipate low tide by taking refuge for a while under a carpet of seaweed or, if the seabed is sandy, inside tunnels where humidity remains constant.

The tidal zone in the Mediterranean, although not comparable in extent and complexity with that to be found along the shores of the Atlantic, nevertheless contains enough material to furnish a complete and unexpectedly colourful aquarium. A rock pool, for instance, may serve as a model in miniature for one type of temperate marine aquarium. Such pools can be seen at low tide in the cavities of rocky coastlines; the water remains permanently, being refilled and renewed regularly by strong waves and the high tide. The organisms that populate these pools are ideal occupants of an aquarium, adapted as they are to a vast range of environmental conditions. Tidal pools, in fact, are continually subjected to the cooling and warming of the surrounding air, attaining temperatures markedly below or above those of the sea water. This occurs not only with the change of seasons, but also within the span of every 24 hours.

Direct observation of such a habitat is the best guide to a choice of organisms, far more numerous than one would expect. In addition to the aforementioned barnacles and limpets, there are sea anemones, sea urchins, starfish, crabs and various species of blennies and gobies, often vividly coloured. Transposing this

The reefs of the Mediterranean are home to innumerable vertebrates, invertebrates and algae. Lobsters, sponges and sea anemones, can all be accommodated in a temperate marine aquarium, provided their subtle associations and relationships are understood and their individual needs respected.

environment into the confines of an aquarium is rewarding if only because it is so easy to collect the different organisms, without the need for elaborate equipment.

The rocky coasts beneath the tidal zone are of even greater fascination and interest. They teem with life, thanks to an incredible variety of microenvironments created by the ceaseless action of waves, currents and living organisms, both plants and animals, large and small. When setting up a marine aquarium inspired by such an underwater habitat, thought must be given, however, to installing a cooling system, because the organisms destined to be its occupants cannot tolerate temperatures above 19° – 20°C (66° – 66°F), particularly for long periods. Once the correct temperature conditions have been established, however, the only restriction to the inventiveness of the aquarist will be the availability of the range of plants and animals. Among the many organisms that can bring interest and colour to a temperate marine aquarium are *Anthias*, wrasse, sea bream, sea bass, sea scorpions, starfish, gorgonians and anemones. The octopus, too, is an ideal occupant of such an aquarium, especially because with this intelligent creature you can establish an almost "human" relationship.

A word of warning, however, to scuba divers and similar underwater enthusiasts. Although it may be tempting to fill your aquarium with species that are literally for the taking, the collecting of marine organisms in this manner is often either forbidden or regulated by law. The type of habitat where the bottom slopes gently and the sediment is increasingly fine, as is the case within calm bays or low coastlines, or at the base of rocks, can provide many an idea for a specialized aquarium. Although apparently uniform and monotonous, sandy or muddy sea bottoms, described as "mobile" in contrast to "rocky," harbour a surprisingly rich fauna. Worms, molluscs and crustaceans, most of which burrow or conceal themselves in the substrate, inhabit these zones in immense numbers. The fish, too, exhibit similar behaviour. In fact, the most characteristic of them – sole, flounder, plaice, and rays – spend much of their time lying in wait for prey, hidden under a thin layer of sand. It becomes clear, therefore, that our initial impression of a zone void of living forms is actually due to the difficulty in detecting the animals, which are so effectively concealed. Yet in addition to those fish species that have developed this means of blending with their surroundings, there are many others that are more conspicuous, notably porgies, some sea bream, giltheads, sea bass, mullet, weevers, gobies, dragonets and pipefish; all of these are acceptable aquarium inhabitants, particularly if young specimens or fairly small species are chosen. These few examples should be sufficient to demonstrate that the temperate marine aquarium can give enormous satisfaction; and if one were to establish a golden rule for setting up such an aquarium, no better sources of inspiration could be sought than the Japanese arts of ikebana and bonsai. Whereas a tropical aquarium appeals for its variety of colour and form, its temperate counterpart, just like these two oriental practices, aims at attention to detail and perfection of taste.

ANOPLARCHUS PURPURESCENS
Fam. Stichaeidae

Common name
Cockscomb.
Dimensions
Up to 20 cm (8 in).
Distribution
From Aleutian Islands to California.
Food
In nature: omnivorous, principally algae, worms and amphipods, but also small molluscs and other crustaceans. In aquarium: shrimp, squid and chopped fish, mixed with fish food containing vegetable matter.
General care
It is a hardy, tranquil species. Use a sandy bottom with stones under which it can hide. Maintain temperatures of 10° – 18°C (50° – 65°F); it can tolerate temperatures up to 20°C (68°F) or more.
Notes
This placid species will breed in captivity. Pairs seek out an area suitable for building a nest in the intertidal zones. The female lays bunches of eggs under stones and watches over them until they hatch within 3 weeks.

APODICHTHYS FLAVIDUS
Fam. Pholididae

Common name
Penpoint gunnel.
Dimensions
Up to 46 cm (18 in).
Distribution
From Kodiak (Alaska) to Santa Barbara (California).
Food
In nature: mainly small crustaceans but also other small invertebrates. In aquarium: shrimp or other forms of chopped seafood mixed with *Artemia* and live invertebrates.
General care
This is a cold-water species which thrives best at below 15°C (59°F) but will adapt to a cool room temperature. Arrange flat stones as shelters on a substrate of sand or gravel. An undergravel filter may be suitable, but it will be necessary to place a screen over the first centimeter of sand to prevent the fish discovering the filter plate. The tank should be low, with a capacity of at least 75 liters (16½ gal).
Notes
Spawning occurs in winter. The eggs are laid in a spherical mass and incubated by one or both parents until they hatch, normally within about 30 days. In captivity the species has been bred and raised up to the stage following larvae metamorphosis.

BALISTES CAROLINENSIS
Fam. Balistidae

Common name
Triggerfish.
Dimensions
Up to 60 cm (24 in) in Atlantic; at least 40 cm (16 in) in Mediterranean.
Distribution
Tropical Atlantic, Mediterranean.
Food
In nature: molluscs, crustaceans and, above all, fish. In aquarium: pieces of fish and squid. Now and then provide whole mussels and crabs.
General care
A temperate water species, it does best at temperatures of 12° – 18°C (53° – 65°F). Accommodate it in tanks of at least 400 liters (88 gal). It tolerates the company of its own kind; do not introduce other fish unless they are of comparable size.
Notes
Found in the vicinity of rocky coasts at depths of 10 – 100 m (33 – 330 ft), this is the only Mediterranean triggerfish. It comes closest to shore in summer, when it spawns. Its colour is brownish. Individuals of inferior rank or in a state of alarm display 3 dark marbled marks.

BLENNIUS OCELLARIS
Fam. Blenniidae

Common name
Butterfly blenny.
Dimensions
Up to 17 cm (6½ in).
Distribution
Mediterranean, eastern Atlantic, English Channel.
Food
Any kind of live food, including meat.
General care
It needs a bed with a thick layer of sand and shells, pots and stones. The aquarium must be dimly lit and the water temperature should be 15° – 21°C (59° – 70°F).
Notes
This is a lively, territorial fish, and as many individuals can be kept together in a tank as there are separate territories available. In fights for division of territory, it is probable that the large dorsal fin, with its black eye spot, plays some part. The species is particularly active during the night. In the wild it spawns in spring. The eggs are laid inside the shells of bivalves and beneath stones, guarded by the males until they hatch.

Anoplarchus purpurescens

Balistes carolinensis

Blennius ocellaris

Apodichthys flavidus

BLENNIUS TENTACULARIS
Fam. Blenniidae

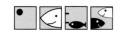

Common name
Horned blenny.
Dimensions
Up to 15 cm (6 in).
Distribution
Mediterranean, Black Sea and Atlantic, from Portugal to Senegal.
Food
In nature: feeds mainly on crustaceans. In aquarium: readily accepts any type of food of animal origin.
General care
Provide a bottom of rocks and pebbles with a little sand, good lighting and water temperatures of 18° – 24°C (65° – 75°F).
Notes
This species lives on seabeds of slime and detritus at depths of up to 30 m (100 ft) and reproduces in spring and summer. The common name derives from the fairly long, branched tentacle above the eye (three times the eye diameter). This tentacle is bigger in males. As a result of recent revised classification, the scientific name is now *Parablennius tentacularis*.

CORIS JULIS
Fam. Labridae

Common name
Rainbow wrasse.
Dimensions
Up to 25 cm (10 in).
Distribution
Mediterranean to Black Sea; eastern Atlantic from Gulf of Guinea to European coasts.
Food
In aquarium: chopped molluscs and crustaceans. Occasionally cleans organic detritus off the bottom.
General care
Needs a soft layer of sand with hiding places, dim lighting and a water temperature of 18° – 21°C (65° – 70°F).
Notes
This fish displays marked changes of colour during growth; this is due to the phenomenon of hermaphroditism, whereby the rainbow wrasse first goes through a long female phase, then a brief transitional phase, and finally a long male phase.

CORYPHOPTERUS NICHOLSI
Fam. Gobiidae

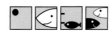

Common name
Black-eyed goby.
Dimensions
Up to 15 cm (6 in).
Distribution
From northern British Columbia to Baia Magdalena (Mexico).
Food
In nature: carnivorous; principally bottom-dwelling crustaceans. In aquarium: chopped seafood such as bivalves, squid, etc.
General care
A sturdy and amenable species. Prepare a sandy bottom with a few stones under which it can hide. It car live in tanks with one or more species, at temperatures of 15° – 22°C (59° – 72°F).
Notes
In nature these gobies spawn from April to October. During this period the male attracts the female into his lair where she lays her eggs, which they both then guard.

DASYATIS PASTINACA
Fam. Dasyatidae

Common name
Common sting ray.
Dimensions
In nature it grows to at least 1.40 m (4½ ft), weighing more than 20 kg (44 lb).
Distribution
Very common in Mediterranean, Black Sea and along Atlantic coasts from Norway to Angola.
Food
Carnivorous: fish, molluscs, crustaceans, etc.
General care
This fish needs a large tank, at least three times the animal's length, with a sandy bottom. Maintain water temperatures of 13° – 25°C (55° – 77°F).
Notes
Take special care when handling these fish because their spines inject poison that may cause tetanus and gangrene; deaths have been recorded.

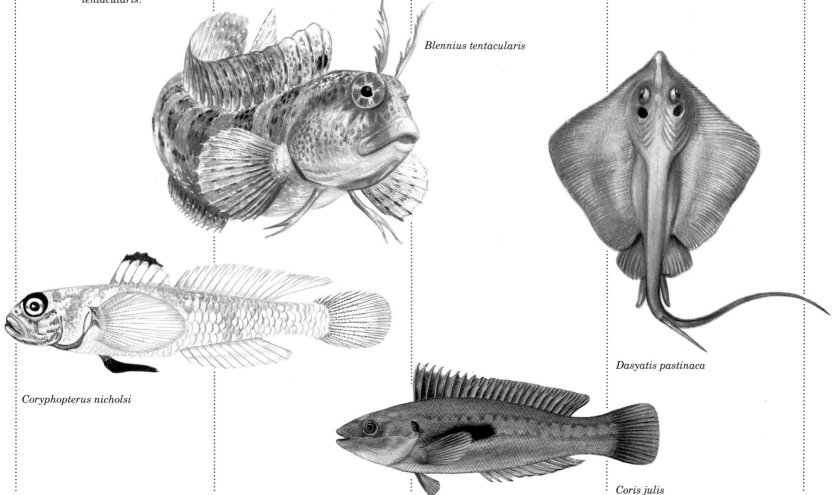

Blennius tentacularis

Coryphopterus nicholsi

Dasyatis pastinaca

Coris julis

TEMPERATE SEA WATERS

EPINEPHELUS GUAZA
Fam. Serranidae

Common name
Mediterranean grouper.
Dimensions
Up to 1.40 m (4½ ft).
Distribution
Eastern Atlantic and Mediterranean.
Food
Carnivorous. In the wild it lies in wait for prey, then launches a lightning attack, biting or swallowing its victim.
General care
The bottom of the tank should be provided with spacious hiding places. Keep lighting dim and maintain temperatures of 16° – 23°C (61° – 74°F).
Notes
Solitary and unapproachable, this fish can only live with individuals larger than itself.

GOBIUS NIGER
Fam. Gobiidae

Common name
Black goby.
Dimensions
Up to 15 – 18 cm (6 – 7 in).
Distribution
Mediterranean to Black Sea, Sea of Azov, Suez Canal; coasts of northeast Atlantic from Cap Blanc (Mauretania) to Norway.
Food
Carnivorous and predatory, it feeds on small live prey, attacking from behind, usually without chasing.
General care
Prepare a soft bottom with hiding places. Maintain water temperatures of 15° – 22°C (59° – 72°F). Tolerates marked fluctuations.
Notes
The black goby is easily raised in the aquarium and will even breed there. The male takes possession of a hiding place, generally below a rock, and cleans it, getting rid of all foreign bodies and carrying them some distance away in his mouth. He then fiercely defends the lair and part of the surrounding territory, inviting the female to lay her eggs and intimidating other males. Given these habits, care must be taken not to put too many individuals in the tank.

HIPPOCAMPUS HUDSONIUS
Fam. Syngnathidae

Common name
Florida seahorse.
Dimensions
Up to 15 cm (6 in).
Distribution
Tropical and temperate zones of western Atlantic, from Florida to New York.
Food
In nature: zooplankton. In aquarium: newly hatched nauplii and live adult *Artemia*. Try to mix with other live crustaceans and small fish such as newborn guppies.
General care
Keep with other quiet species such as pipefish, or alone with individuals of the same species. Put vegetation or fine coral into the tank, to which they can attach themselves.
Notes
This species comes in a large variety of colours such as grey, brown and red. It is relatively easy to keep and has been bred in captivity.

JORDANIA ZONOPE
Fam. Cottidae

Common name
Longfin sculpin.
Dimensions
Up to 15 cm (6 in).
Distribution
From Alaska to central California.
Food
In nature: small crustaceans and other benthic animals. In aquarium: live *Artemia*, chopped shrimp and other crustaceans.
General care
Thrives best in an aquarium at a temperature of below 10°C (50°F) but adapts to those of up to 20°C (68°F). Use large rocks with flat vertical surfaces so that they can settle on them and make cracks for hiding.
Notes
Fertilization is internal. The eggs are normally laid in small bunches. The young are relatively big and easy to raise.

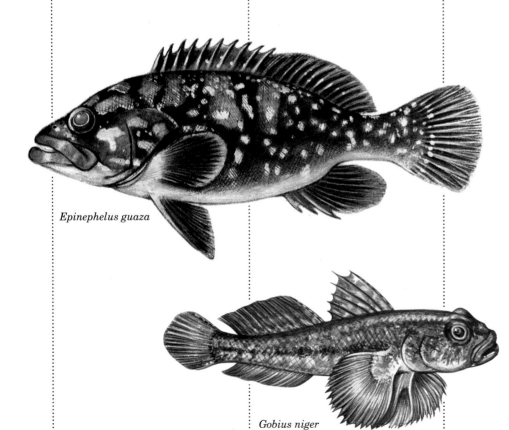

Epinephelus guaza

Gobius niger

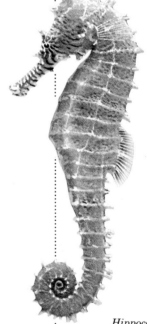

Hippocampus hudsonius

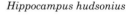

Jordania zonope

LABRUS VIRIDIS
Fam. Labridae

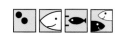

Common name
Green wrasse.
Dimensions
Up to 45 cm (18 in).
Distribution
Eastern Atlantic and Mediterranean.
Food
Carnivorous: chopped fish and crustaceans. It is easy to feed.
General care
This wrasse needs a tank with plenty of hiding places and a great deal of free swimming space, good lighting and water temperatures of 18° – 21°C (65° – 70°F).
Notes
Individuals measuring under 27 cm (10½ in) are females, while those of more than 38 cm (15 in) are males. Livery is fairly variable, ranging from green to brown with deviations to white, yellow, red and orange. However, contrary to what happens among other Labridae, change of sex does not bring about change of colour.

MURAENA HELENA
Fam. Muraenidae

Common name
Moray eel.
Dimensions
Up to 1 – 1.5 m (3 – 5 ft).
Distribution
Mediterranean and subtropical shores of eastern Atlantic.
Food
Feeds mainly on invertebrates, especially cuttlefish, octopus and shrimp.
General care
Because in the sea it hides in caves and rock crevices, it needs artificial places of concealment in the aquarium (e.g. large pots and jars). Lighting should be dim and the water temperature 18° – 21°C (65° – 70°F).
Notes
Moray eels grow to a considerable size so they are most suitable for public aquaria, but it is possible to accommodate single specimens in a home aquarium, even in conjunction with other selected species.

NAUTICHTHYS OCULOFASCIATUS
Fam. Cottidae

Common name
Sailfin sculpin.
Dimensions
Up to 20 cm (8 in).
Distribution
From Alaska to central California.
Food
In nature: carnivorous, feeding mainly on crustaceans. In aquarium: live *Artemia*, chopped shrimp and squid.
General care
This is a marvellous cold-water aquarium fish. Create fairly large cavities and keep several individuals together at temperatures of 10° – 13°C (50° – 55°F).
Notes
The fish are most active at night, remaining by day in small caves, often hanging upside down from the roof. The large anterior dorsal fin is used as a primitive bait to attract crustaceans. In the presence of likely prey, it is waved back and forth; the fish then lowers it to mouth level and catches any unwary victim which may have strayed too close.

OPSANUS TAU
Fam. Batrachoididae

Common name
Atlantic toadfish.
Dimensions
Up to 25 cm (10 in).
Distribution
From Maine to Florida.
Food
In nature: carnivorous. In aquarium: whole or chopped fish, squid, etc.
General care
Easy to keep; avoid putting it in the aquarium with smaller fish that may be eaten. Create caves or introduce pipes and pots for egg laying. Maintain temperatures of 18° – 24°C (65° – 75°F).
Notes
This species is notable for the strange sounds it makes during courtship, consisting of grunts, hums, croaks and whistles. The eggs are laid in shallow water, often inside cans, shells or other natural shelters. The male watches over the eggs for 10 – 25 days, until they hatch.

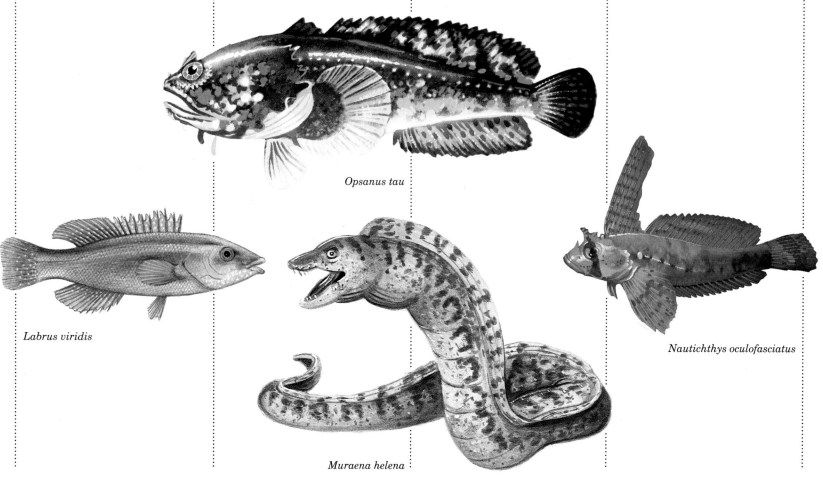

Opsanus tau

Labrus viridis

Muraena helena

Nautichthys oculofasciatus

TEMPERATE SEA WATERS

PLEURONICHTHYS COENOSUS
Fam. Pleuronectidae

Common name
Turbot.
Dimensions
Up to 36 cm (14½ in).
Distribution
From Bay of California to Alaska.
Food
In nature: carnivorous. In aquarium: a wide variety of food, including chopped fish and other types of seafood.
General care
It is a hardy species but should not be kept with other aggressive fish. The bottom of the tank needs to be covered with sand and the temperature kept at 10° – 24°C (50° – 75°F).
Notes
Only small individuals are suitable for home aquaria. They are common in bays and can be easily caught. They are similar in habit to the flounder (*Platichthys flesus*) of the Mediterranean, which is capable of adapting to brackish and even fresh water. In the aquarium young individuals acclimate best; these are to be found more readily during the autumn when they move in from the sea into lagoons and close to river mouths in search of less salty water.

PRISTIGENYS SERRULA
Fam. Priacanthidae

Common name
Popeye catalufa.
Dimensions
Up to 33 cm (13 in).
Distribution
From southern California to Peru.
Food
In nature: carnivorous. In aquarium: chopped fish, shrimp, squid, bivalves.
General care
Make dark holes as hiding places. It is a tranquil species, suitable for living with other varieties. Keep water temperatures of 18° – 22°C (65° – 72°F).
Notes
These colourful fish are excellent for the aquarium, once acclimated. Although they have nocturnal habits, they soon learn to feed by day. They inhabit fairly deep water, normally down to 30 m (100 ft), both on coral reefs and above sandy beds. Sometimes they are seen by scuba divers, and they may be caught accidentally by boats trawling for shrimp.

RAJA CLAVATA
Fam. Rajidae

Common name
Thornback ray.
Dimensions
Up to about 1 m (3 ft).
Distribution
Eastern Atlantic and Mediterranean.
Food
Carnivorous; small fish, molluscs, chopped crustaceans, etc.
General care
Needs a fairly large tank (at least three times the animal's length). The bottom must be generously covered with sand because the fish is in the habit of burrowing, lifting itself occasionally to swim with flapping movements of the pectoral fins, i.e. the disc margins. The water temperature should be 15° – 25°C (59° – 77°F), but otherwise it tolerates a variety of conditions.

RHAMPHOCOTTUS RICHARDSONI
Fam. Cottidae

Common name
Grunt sculpin.
Dimensions
Up to about 7 cm (2¾ in).
Distribution
From Alaska to central California.
Food
In nature: small benthic, bottom-swimming crustaceans. In aquarium: adult *Artemia* and other live crustaceans, chopped shrimp and squid.
General care
Create a rocky environment with caves, holes and crevices. Keep in a tank of 75 liters (16½ gal) or more with one or more tranquil species. It prefers temperatures under 15°C (59°F) but will tolerate up to 23°C (74°F).
Notes
This species lives mainly along rocky coasts, its habitat ranging from rock pools to water up to some 30 m (100 ft) deep; it has also been found on sandy bottoms and at a depth of 165 m (540 ft). It uses its pectoral fin for resting and crawling but it can also swim normally. As a rule it spawns in winter, laying small quantities of yellow or orange eggs.

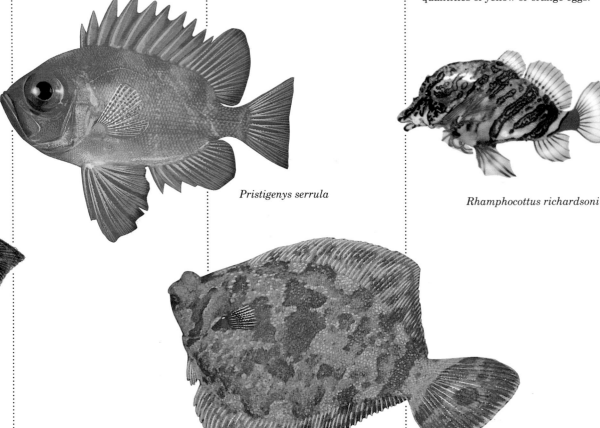

Pristigenys serrula

Rhamphocottus richardsoni

Raja clavata

Pleuronichthys coenosus

SCORPAENA GUTTATA Fam. Scorpaenidae	SCORPAENA SCROFA Fam. Scorpaenidae	SCYLIORHINUS STELLARIS Fam. Scyliorhinidae	SCYTALINA CERDALE Fam. Scytalinidae
		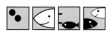	

Common name
California scorpion fish.
Dimensions
Up to 43 cm (17 in).
Distribution
From central California to Gulf of California.
Food
In nature: carnivorous; it is an opportunistic fish which preys on a wide variety of benthic organisms. In aquarium: fish, chopped squid, etc.
General care
As a rule only small specimens are kept in home aquaria. It is a sedentary species which does not need much space. Only young individuals, because of their limited capacity for predation, can live together with other species, and these must be larger than themselves. Maintain temperatures of 15° – 22°C (59° – 72°F).
Notes
Unlike most other members of the family Scorpaenidae, *S. guttata* lays eggs wrapped inside a transparent pear-shaped sac. Beware of the spines on the body as they are poisonous.

Common name
Red scorpion fish.
Dimensions
Up to 30 cm (12 in).
Distribution
Eastern Atlantic and Mediterranean.
Food
Since it is a carnivore and a predator, its aquarium diet presents no difficulties, provided all meaty food is soft.
General care
It needs a bottom with groups of stones and hiding places. The water temperature should be 17° – 24°C (63° – 75°F).
Notes
The spiny rays of the dorsal fin have poisonous glands at the base; injuries can be very painful.

Common name
Nursehound, dogfish.
Dimensions
Up to 80 cm (32 in).
Distribution
Eastern Atlantic and Mediterranean.
Food
Fish, molluscs and marine invertebrates.
General care
It needs a tank at least three times its length. Maintain water temperatures of 15° – 25°C (59° – 77°F).
Notes
The nursehound is mainly nocturnal by habit. In nature it lays characteristic eggs, each being enveloped in a horny shell, with four long, twisting filaments which adhere to submerged organisms, especially *Gorgonia* species. These fish are shark-like and are often seen in public aquaria, where they live contentedly.

Common name
Graveldiver.
Dimensions
Up to 15 cm (6 in).
Distribution
From southeast Alaska to central California.
Food
In nature: interstitial organisms. In aquarium: nauplii and adult *Artemia*, worms, shrimp and small chopped crustaceans.
General care
Keep in a small 40 – liter (9 – gal) tank. Use an undergravel filter with a screen to prevent the fish finding their way underneath. Cover the screen with a few centimeters of gravel. Temperatures should be 5° – 10°C (41° – 50°F).

Scytalina cerale

Scorpaena guttata

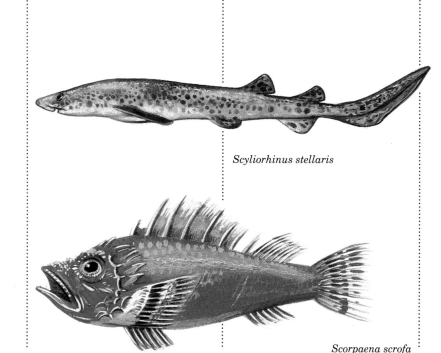

Scyliorhinus stellaris

Scorpaena scrofa

SEBASTES ROSACEUS
Fam. Scorpaenidae

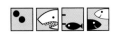

Common name
Rosy rockfish.
Dimensions
Up to 36 cm (14½ in).
Distribution
From California to peninsula of central California, Mexico.
Food
In nature: carnivorous. In aquarium: chopped fish, shrimp and squid.
General care
Suitable for keeping in tanks with one or more species. Adults need an aquarium of 1,000 liters (220 gal) or more with numerous rock-like shelters along the back and at one end of the tank, plus plenty of space for swimming. Several specimens can be kept together. Temperatures should be 15° – 20°C (59° – 68°F).
Notes
This fish is often caught by sporting fishermen, usually at a depth of more than 15 m (50 ft). It is necessary to degas the swim bladder with a hypodermic needle or to subject the fish to decompression. The few specimens that survive capture at the surface with rod and line make excellent aquarium fish.

SEBASTES RUBRIVINCTUS
Fam. Scorpaenidae

Common name
Yelloweye rockfish.
Dimensions
Up to 36 cm (14½ in).
Distribution
Southern California, peninsula of northern California, Mexico.
Food
As for *Sebastes rosaceus*.
General care
Prepare rocks for shelter. Adults need an aquarium of 1,000 liters (220 gal) or more. Recommended temperatures are below 15°C (59°F) but the fish may adapt to around 23°C (74°F).
Notes
The dorsal and anal fins are provided with spiny rays linked to poison glands with a slightly toxic secretion. For this reason, it is good procedure, as with all Scorpaenidae, particularly the tropical species, to be very careful when handling the fish. *S. rubrivinctus* has a liking for rocky zones surrounded by beds of sand, mud or detritus. The species *S. serciceps* lives in shallower water and is easily caught by scuba divers; it is necessary to degas the swim bladder as soon as the fish is taken. The colours make it highly attractive for the aquarium and it soon adapts to life in captivity.

SERRANUS CABRILLA
Fam. Serranidae

Common name
Sea bass.
Dimensions
Up to 25 cm (10 in).
Distribution
Mediterranean, Atlantic from British Isles to Angola, North Sea, Red Sea.
Food
In the wild it feeds mainly on small fish, but also crustaceans.
General care
It adapts very well to aquarium life. Prepare a bed with crevices and hiding places as in nature it lives among rocks covered with algae in the upper coastal zone. It needs moderate light and a water temperature of 18° – 21°C (65° – 70°F).
Notes
The species is frequently encountered in the Mediterranean, for it is inquisitive and often approaches divers spontaneously. It occupies a wide variety of habitats (reefs, beds of detritus, fields of *Posidonia*) and will range from the surface down to a depth of 500 m (1,650 ft).

SYNGNATHUS LEPTORHYNCHUS
Fam. Syngnathidae

Common name
Pipefish.
Dimensions
Up to 24 cm (9½ in).
Distribution
From Alaska to peninsula of southern California, Mexico.
Food
In nature: carnivorous, principally small crustaceans including amphipods, copepods and crab larvae. In aquarium: live *Artemia*, adults and nauplii, and other live zooplankton.
General care
Keep only with non-aggressive species or others of its own kind in tanks of 75 liters (16½ gal) or more. Recommended temperatures are 10° – 22°C (50° – 72°F).
Notes
There are several Mediterranean species of pipefish (*S. phlegon*, *S. acus*, *S. typle*, etc.), notable for their long, slender bodies, that are possible occupants of a temperate marine aquarium. Although once common, they have now become quite rare due to pollution and alteration of the seabed.

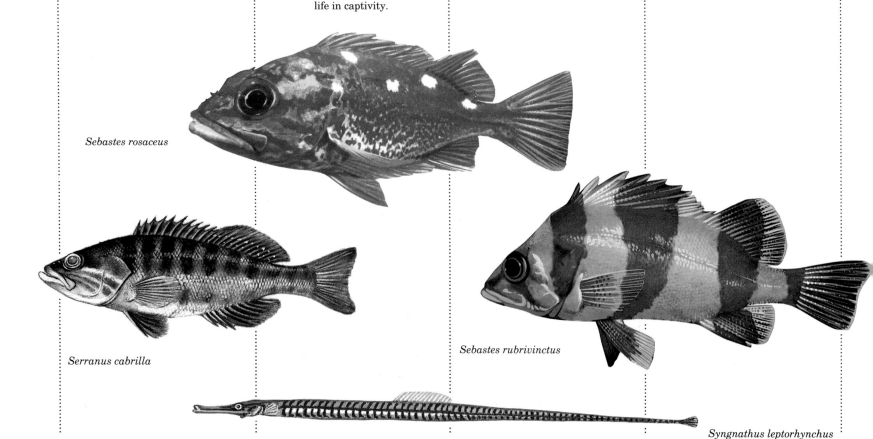

Sebastes rosaceus

Serranus cabrilla

Sebastes rubrivinctus

Syngnathus leptorhynchus

THALASSOMA PAVO
Fam. Labridae

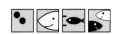

Common name
Peacock wrasse.
Dimensions
Up to 20 cm (8 in).
Distribution
Eastern Atlantic and Mediterranean.
Food
In nature: mainly carnivorous. In aquarium: thrives well on chopped seafood and other normal aquarium diets.
General care
As a rule, it adapts well to an aquarium with other species, but it may turn aggressive when it reaches its maximum size. Cover the bottom with sand in which the fish can burrow at night. Maintain temperatures of 18° – 24°C (65° – 75°F).
Notes
This is the most colourful of the Mediterranean Labridae and the one that prefers warmer water, so that it can be kept in a tropical aquarium. Individuals may exhibit a progressive change of sex. Smaller and relatively less colourful specimens are females, while larger, more vividly coloured specimens are males.

TORPEDO MARMORATA
Fam. Torpedinidae

Common name
Marbled torpedo ray.
Dimensions
Up to 60 cm (24 in).
Distribution
Eastern Atlantic and Mediterranean.
Food
Carnivorous; small fish, molluscs, chopped crustaceans, etc.
General care
This ray needs a large tank, the bottom of which should be covered with plenty of sand, where it will prefer to live. Maintain water temperatures of 13° – 25°C (55° – 77°F).
Notes
The electric shocks delivered by torpedo rays are not dangerous to humans.

TRIPTERYGION NASUS
Fam. Tripterygiidae

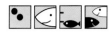

Common name
Blenny.
Dimensions
Up to 7 cm (2¾ in).
Distribution
Mediterranean and Adriatic.
Food
In nature: mainly bottom-dwelling crustaceans. In aquarium: live *Artemia* and other planktonic organisms, small worms, fish food.
General care
Thrives best with individuals of its own kind but can be kept with other peaceful species. Create a substrate with large stones. Temperatures should be 22° – 25°C (72° – 77°F).
Notes
The male assumes bright colours when spawning; the females are brown. These fish live in rocky zones and can be found down to a depth of 25 m (80 ft).

ZALIEUTES ELATER
Fam. Ogcocephalidae

Common name
Cortez batfish.
Dimensions
Up to 15 cm (6 in).
Distribution
From southern California to Peru.
Food
In nature: carnivorous; fish, crustaceans, worms and molluscs. In aquarium: live *Artemia* and shrimp, bivalves and chopped squid.
General care
An individual of 10 cm (4 in) needs a tank of at least 100 liters (22 gal), a bed of fine sand, dim lighting and a water temperature of 18° – 21°C (65° – 70°F).
Notes
Members of this and related genera inhabit tropical and subtropical regions of the Atlantic, Pacific and Indian Oceans. Most of them are caught accidentally by trawlers at depths of more than 100 m (330 ft). Some, however, such as *Ogcephalus radiatus*, live at around 10 m (33 ft). Also related to *Zalieutes* are *Halieutaea fumosa*, *H. retifera* and *Dibranchus erythrinus*.

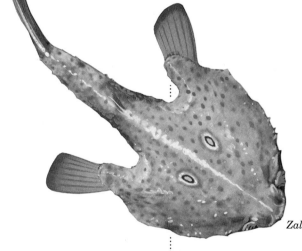

Zalieutes elater

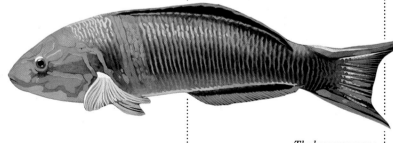

Thalassoma pavo

Torpedo marmorata

Tripterygion nasus

PLANTS AND INVERTEBRATES

Plants and invertebrates

Invertebrates (bottom) can either complement an aquarium or form a distinctive collection on their own. Sometimes they live quite happily with other organisms (below).

PLANTS

Plants and invertebrates are in many cases natural complements to an aquarium. Sometimes, however, they constitute the very theme of a tank. These aquaria are designed to accommodate only plants, harmoniously arranged according to their dimensions, leaf colouring and methods of growth in order to create delightful underwater gardens.

Yet the function of plants in the aquarium is by no means merely aesthetic. On the contrary, thanks to photosynthesis, plants replenish the aquarium with pure oxygen during the hours of daylight, even though in darkness they consume it and produce carbon dioxide. Apart from this, plants utilize and decompose organic substances, serving as effective filtrating materials. They also provide shelter to fry and the more timid fish. Additionally, they can be transformed into an ideal substrate where many species of fish can lay their eggs, and they can be mixed with other foods to constitute an excellent supplement to the aquarium diet.

Lack of scientific precision sometimes defines all underwater vegetation generically as algae, but this is an error. In addition to algae proper, there are true plants furnished with roots, stems, leaves, flowers and fruit. In fact, the principal difference between algae and plants is one of structure rather than form, which is often deceptive. Algae are divided into three main groups, green, red and brown, according to their predominant photosynthetic pigments, plus a number of minor groups, consisting of unicellular or microscopic organisms. All lack, by definition, an anatomical structure with recognizably different tissues or organized systems, as is the case with higher plants. Nevertheless, and especially in larger algae, such as certain brown forms (*Laminaria*, *Macrocystis*) which may grow to many feet in length, it is possible to discern structures that, outwardly at least, may be mistaken for roots, stems and leaves. Despite appearances, however, these are invariably morphological adaptations of the thallus, the term normally used to describe the entire structure of an alga.

Aquatic plants, on the other hand, have the same anatomical–functional structures as land plants (roots, stems, leaves, etc.), although they differ in a number of ways, most importantly in the lack of stiff supporting elements. The higher density of water, in fact, allows submerged plants to grow straight up, without resorting to strong tissues such as wood, or to structures comparable to trunks and branches. This is the same kind of adaptation to the environment as applies to a whale, whose body can grow and be supported solely by the surrounding water, whereas the gigantic animal is crushed under its own weight when stranded ashore. This fact is best demonstrated when a true aquatic plant is plucked from its proper environment; the same organism that possesses an upright, ramified form when submerged, collapses when exposed to the air, retaining nothing of its original appearance.

Another factor that differentiates algae and aquatic plants from terrestrial vegetation is the absence of protective membranes designed to prevent the loss of water and consequent dehydration of the cells. It may seem a minimal difference but actually it enables aquatic plants to absorb organic and inorganic compounds, not simply through roots but through each and every one of their parts.

Living underwater, however, is not wholly advantageous. The major problem facing aquatic plants is obtaining sufficient light energy to bring about the chemical reactions associated with photosynthesis, i.e. the transformation of inorganic compounds into more complex organic substances. The same characteristics that in many ways make water an ideal environment impose restrictions as far as light is

Colonies of soft corals and starfish are among the most beautiful examples of the many thousands of species of invertebrate.

concerned. Water, in fact, is a liquid that markedly hinders light from penetrating to any great depth. A ray of sun striking the surface of the sea or a lake is in large measure reflected as if it were hitting a mirror, so that only a small fraction of its original energy will penetrate below the waves. Under the surface its path may be obstructed by phenomena of refraction and diffraction due to suspended sediments and the very organisms present in the water. Even more importantly, as it travels downward, light is progressively deprived of certain characteristic wavelengths. White light is made up of a number of coloured radiations, as is revealed by a rainbow or a crystal prism. The various colours thus obtained by the decomposition of sunlight each possess a different wavelength, i.e. a different energy. Confronted by light, water acts as a filter that, according to depth, progressively selects the various chromatic components of an individual ray. If we were to test the principle by immersing a board painted with colours of the rainbow, we would notice that after a few feet the red would start to fade until it appeared to be black. The same happens gradually to orange and yellow, to the point that eventually the warm colours are no longer visible and the only distinguishable tones are green and blue. This progressive disappearance of colours, which depends on the gradual diminution of light energy, has its effect on plants, for their photosynthetic pigments (chlorophyll) are sensitive or adapted to absorbing specific parts of the spectrum, exhibiting a preference for the wavelengths that correspond to red, known as phytostimulants. Some groups of algae, however, are adapted to the absorption of wavelengths corresponding to green, yellow or blue.

For this reason the ocean depths, to a far greater extent than bodies of fresh water, exhibit a completely different vertical distribution of vegetation, in which the green algae and flowering plants (*Posidonia* and *Zostera*) are concentrated in the upper zones, the brown algae predominate in the intermediate zones and the red algae extend into the depths.

It should be evident from all this that proper lighting of a freshwater, and especially a marine, aquarium is of crucial importance, requiring the assistance of special "photosynthetic" lamps. As a rule, best results are obtained by using lamps with different characteristics. Moreover, it must be borne in mind that the lighting requirements of aquarium plants often differ from one species to another.

INVERTEBRATES

The role of invertebrates in an aquarium, particularly in the marine tank where the majority are likely to be accommodated, is very different from that of plants and fish, although, since they are animals, some appear to have more in common with the latter. Fundamentally they can enrich the aquarium, either by complementing the whole or by being displayed singularly on their own merits. A tank containing only hard corals, tropical *Actinia* (sea anemone) and other invertebrates can prove just as fascinating as one containing various coloured fish. In other cases their presence may serve to complete a thematic aquarium or be necessary for the proper maintenance of certain fish species, such as the clownfish in their symbiotic association with the sea anemones.

A comprehensive introduction to the complex world of invertebrates would require a book to itself, with lengthy chapters devoted to individual groups. Nor is it feasible to make brief summaries of the available material in the pattern of our introductory sections to fish. The dictionary definition of an invertebrate is simple enough: "An animal without a bony skeleton such as characterizes a vertebrate." But their division into types, classes, orders, families, not to mention species and varieties, is even today, in many cases, a topic of controversy among the specialists. The basic reason for the problems in classifying invertebrates is their sheer abundance. They are to be found everywhere, both in fresh and sea water, in tropical, temperate and polar regions. In virtually all watery environments there are representatives, to varying degrees, of every single group, often in numbers and densities that almost defy belief, as for instance in and around the enormous Great Barrier Reef, which covers an area of over 200,000 km^2 (77,000 sq. miles). Resorting to numbers is probably the only way to convey any idea of the vast complexity of the world of aquatic invertebrates. For example, in the Indo-Pacific region alone there are 500 species of corals and 5,000 species of molluscs. According to latest research, there are some 2,000 species of starfish and more than 10,000 sponges. The number of crab and crayfish species is no less than 39,000, while the total of freshwater and saltwater annelid worm species is in excess of 11,000. Included among the invertebrates are several living fossils, such as *Neopilina*, discovered in 1952, *Limulus*, closely related to the extinct trilobites, and *Nautilus*, descended from the Mesozoic ammonites. Given an evolution that originated so far back in time, it is understandable that present-day invertebrates exhibit all manner of reproductive methods, feeding habits and behaviour patterns, plus an immense variety of adaptations and specializations designed for survival in one or another of the numerous micro- and macrohabitats that exist in a watery world. It is precisely in this respect that the aquarist's opportunities are limited. The aquarium environment, however perfect, will always be restricted and inadequate in comparison with the natural world outside. So the aquarist's choice will be circumscribed not only because of their own capacity but also because of their need to respect the biological and behavioural characteristics of the organisms in their care, whether they be invertebrates, plants or fish.

APONOGETON CRISPUS
Fam. Aponogetonaceae

Dimensions
Leaves up to 30 cm (12 in) long.
Distribution
Southeast Asia, Sri Lanka. It is found in enclosed watery environments subject to drying out during periods of drought.
General care
Fairly robust plant, needing bright light and a water temperature of 20° – 28°C (68° – 82°F). Feed with peat and add clay to the soil. An adequate rest period is essential (see *Notes*).
Notes
This plant produces a series of submerged leaves, lanceolate with slightly undulate margins, grouped in rosettes. The appearance of floating leaves may suggest the presence of a plant hybridized with other species. This *Aponogeton* is suitable for planting in the center and at the sides of the aquarium. It is particularly important when growing all *Aponogeton* species to give them a rest period. When the leaves, having reached their maximum size, begin to fade, transfer the plants to an unheated aquarium for about two months before transplanting them again to the original tank. Such operations are made easier by keeping these species in buried pots.

CABOMBA AQUATICA
Fam. Cabombaceae

Dimensions
Leaves up to 7 cm (2¾ in) long.
Distribution
Tropical America and Amazon region. It grows naturally in shallow water that is calm or with a gentle current.
General care
Needs bright to very bright light and slightly acid, well aerated water. Add clay to the soil. Maintain temperatures of 22° – 26°C (72° – 79°F).
Notes
This is considered one of the loveliest plants for the tropical freshwater aquarium. It lives permanently submerged and grows from a rhizome anchored to the bottom by very thin roots. The upright stem bears finely divided submerged leaves, while the surface leaves are smaller and less divided. The flowers, on a long stalk, are white or yellowish. It is reproduced by cuttings. Similar in appearance is *Cabomba caroliniana*, less demanding in respect of temperature, a variety of which, *C. pulcherrima*, has typical reddish leaves.

CAULERPA TAXIFOLIA
Fam. Caulerpaceae

Dimensions
Leaves up to 10 – 20 cm (4 – 8 in) long.
Distribution
Tropical Atlantic, Red Sea, Indo-Pacific. Recently observed in the Mediterranean.
General care
Needs good lighting and strong water movement.
Notes
Caulerpa taxifolia is the most common species in home aquaria. The various species belonging to the genus *Caulerpa* (*C. taxifolia*, *C. cupressoides*, *C. serrulata*, *C. verticillata*, etc.) are all easy to cultivate in the aquarium. It should be remembered that if these exotic species are placed in ordinary tap water, they may react in unexpected ways. For temperate marine aquaria, the Mediterranean species *C. prolifera*, with long leaves about 2 cm (¾ in) wide, is recommended.

CRYPTOCORINE AFFINIS
Fam. Araceae

Dimensions
Leaves up to 40 cm (16 in) long.
Distribution
Malaysian peninsula; frequently found in swamps or zones with weak currents where it forms dense masses.
General care
Needs dim and very constant lighting, which, if too strong, will cause the leaves to fall. Maintain temperatures of 22° – 24°C (72° – 75°F).
Notes
This is one of the most popular aquarium plants because it is so easy to grow. In the tank the lanceolate leaves take on an emerald green colour above, while the undersides have reddish tints. It is easily propagated by stolons and within quite a short time the vegetation is dense. Among other well-known species of the same genus, *Cryptocorine balansae* has soft, finely undulate leaves, *C. ciliata* has long leaves that may protrude from the tank and *C. nevilli* has small leaves of 5 cm (2 in) that can form a thick carpet over the bottom of the aquarium.

ECHINODORUS CORDIFOLIUS
Fam. Alismataceae

Dimensions
Leaves up to 20 cm (8 in) long and 15 – 20 cm (6 – 8 in) wide.
Distribution
Southeastern part of North America, Mexico and West Indies. The species is typical of swampland, growing partially above the surface in zones subject to variations in water level.
General care
A fairly sturdy species, needing strong and prolonged lighting. This helps the emergent leaves to develop, such growth being limited if the period of illumination is less than 12 hours daily. Young plants, however, need about 16 hours of light.
Notes
The dimensions of this species of *Echinodorus* make it suitable to be grown on its own and in large uncovered aquaria. Other species equally useful in the aquarium are *E. major* and *E. bleheri*, both of large size. In a 200-liter (44-gal) tank it is possible to plant and keep in groups *E. amazonicus*, with lanceolate leaves, and *E. tenellus*, which quickly forms dense prairies.

ELODEA CANADENSIS
Fam. Hydrocharitaceae

Dimensions
The stems can grow to a length of several feet.
Distribution
Originally from North America, it was introduced to Ireland and then spread to the whole of Europe. It prefers swampy zones and water with weak currents.
General care
A fairly robust species, highly suitable for the temperate aquarium. It needs hard water and good lighting.
Notes
This species grows freely in swamps and around springs, but before introducing it to the aquarium, it is advisable to give it a disinfectant dip in salt water to get rid of parasites. It soon forms a thick carpet and is propagated easily by simple stem division. Strangely, only female specimens of the species exist, descended from the plants imported into Ireland in 1836. Commonly known as pondweed, it spread rapidly to other areas. Other members of the family include *Egeria densa*, *Lagarosiphon major* and *Hydrocharis morsus-ranae*. The last floats on the water surface where its rosettes of leaves form thick carpets. All the Hydrochatitaceae are suited to cold-water aquaria.

MYRIOPHYLLUM SPICATUM
Fam. Haloragaceae

Dimensions
Leaves up to 3 – 4 cm (1¼ – 1½ in) long. Stem 30 – 50 cm (12 – 20 in) high (several feet in nature).
Distribution
Temperate European waters, still or flowing. Also adapts to cloudy, nutrient-rich water.
General care
Suitable for temperate aquaria with temperatures of 12˚ – 20˚C (54˚ – 68˚F). Does not need strong lighting. Prefers hard, alkaline water. Cannot tolerate any materials in suspension that may be deposited on its leaves, slowing growth and spoiling appearance.
Notes
The genus *Myriophyllum* owes its name to the numerous pectinate-pinnate leaves that give the plant its feathery appearance. The colour is bright green but certain species, if exposed to strong light, may take on reddish tints. In tropical aquaria it is possible to cultivate exotic species such as *M. mattogrossense*, with reddish foliage, and *M. brasiliense*, emerald green and of compact appearance, which needs good lighting and tends to grow above the surface.

VALLISNERIA SPIRALIS
Fam. Hydrocharitaceae

Dimensions
Leaves up to 50 cm (20 in) and more in length.
Distribution
Southern Europe and northern parts of Africa. Introduced to many tropical and subtropical zones. Prefers still or weakly flowing water.
General care
Sturdy species without any particular requirements concerning the chemical properties of the water. Needs good lighting. Adapts to a wide range of temperatures, from 12˚ – 30˚C (54˚ – 86˚F). Add clay to soil.
Notes
This is the best known species of *Vallisneria*, with ribbon-like leaves in rosettes. It is easily propagated by stolons, quickly covering the available surface of the tank, and is recommended above all for masking the tank bottom. The variety *V. spiralis tortifolia* is popular and very suitable for a home aquarium, its fairly small leaves, 20 cm (8 in) long, growing spirally. The species *V. gigantea* has longer and broader leaves, and is better suited to larger and more tropical aquaria.

ACROPORA CERVICORNIS
Cl. Anthozoa

Dimensions
Up to 3 m (10 ft) tall.
Distribution
Florida Keys, Bahamas and West Indies, to Brazil.
Food
In nature: plankton. In aquarium: *Artemia* nauplii.
Notes
This is a difficult species to keep in a live state and not really to be recommended. The skeleton can serve as an aquarium ornament and a refuge for the fish, although trade in this hard coral should not be encouraged.

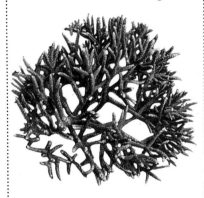

ACTINIA EQUINA
Cl. Anthozoa

Dimensions
Diameter 6 – 7 cm (2½ – 2¾ in).
Distribution
Mediterranean, European Atlantic coasts to English Channel and North Sea.
Food
In nature: planktonic organisms and small fish. In aquarium: pieces of mollusc or fish.
Notes
Commonly found on rocks at the surface and in rock pools. Adapts very well to aquarium life in an unheated tank, where initially it will move from place to place, seeking an ideal spot to settle.

ANTHOPLEURA XANTHO-GRAMMICA   Cl. Anthozoa

Dimensions
Up to 20 cm (8 in) in diameter.
Distribution
North Pacific from Alaska to Panama.
Food
In nature: fish and other sea organisms. In aquarium: whole or chopped seafood placed near tentacles.
Notes
This large sea anemone and the related, smaller *A. elegantissima*, from the North Pacific, should be kept at around 10°C (50°F), in a cold marine aquarium. Specimens caught in warmer regions should be kept at average temperatures.

ARBACIA PUNCTULATA
Cl. Echinoidea

Dimensions
Up to 5 cm (2 in) in diameter.
Distribution
Mediterranean and tropical eastern Atlantic.
Food
In nature: mainly herbivorous. In aquarium: principally algae. May also behave as a cleaner species.
Notes
Species of this genus are generally used in scientific laboratories for research in embryology and genetics.

ASTERIAS FORBESI
Cl. Asteroidea

Dimensions
Up to 13 cm (5¼ in).
Distribution
From Gulf of Maine to Texas.
Food
In nature: principally bivalves, mussels and oysters. In aquarium: as in nature.
Notes
These starfish are frequently found around piers and jetties along the Atlantic coast of North America.

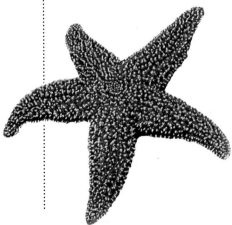

ASTERIAS RUBENS
Cl. Asteroidea

Dimensions
Up to 50 cm (20 in).
Distribution
Atlantic, North Sea, Baltic Sea.
Food
Feeds mainly on molluscs. In some areas it can do much damage to oyster beds.
Notes
Similar to this species is the Mediterranean red starfish which is found near the surface and which adapts well to aquarium life.

BALANOPHYLLA ELEGANS
Cl. Anthozoa

Dimensions
Up to 12 cm (5 in) tall.
Distribution
British Colombia to Mexico.
Food
In nature: plankton. In aquarium: does well on *Artemia* nauplii and adults, but also other live planktonic organisms.
Notes
B. elegans should be kept in a cool aquarium at temperatures of 10° – 20°C (50° – 68°F), with dim lighting.

CERIANTHUS sp.
Cl. Anthozoa

Dimensions
The larger species may have a tube of over 60 cm (24 in) in height and a tentacle aperture of 15 cm (6 in).
Distribution
Warm and cold seas all over the world.
Food
In nature: preys on various marine organisms. In aquarium: whole sea organisms, if small, or in pieces, placed on tentacles.
Notes
The *Cerianthus* species need a deep substrate for burrowing. The ideal temperature depends on the species and its natural habitat.

CONUS CALIFORNICUS
Cl. Gastropoda

Dimensions
Up to 5 cm (2 in).
Distribution
Warm seas, from central California to Lower California (Mexico).
Food
In nature: preys on fish, worms, molluscs and small sea animals. In aquarium: live fish, worms and other marine animals.
Notes
The Conidae possess a strong poison which is injected through the teeth and which may be fatal to humans, but this does not appear to be the case with *C. californicus*. The majority of species are not recommended for home aquaria.

CORYNACTIS CALIFORNICA
Cl. Anthozoa

Dimensions
Up to 2.5 cm (1 in) in diameter.
Distribution
From central California to Mexico.
Food
In nature: plankton. In aquarium: live *Artemia* nauplii and adults, other small crustaceans and finely chopped seafood.
Notes
This bright and attractive sea anemone is a colourful addition to a temperate marine aquarium. It adapts well to temperatures of 20°C (68°F) or below.

CROSSASTER PAPPOSUS
Cl. Asteroidea

Dimensions
Up to 18 cm (7 in).
Distribution
Arctic Ocean to Gulf of Maine; Alaska to state of Washington.
Food
In nature: preys on other starfish.
Notes
This is one of the most beautiful starfish. It needs cold water, at temperatures of below 13°C (55°F).

CUCUMARIA MINIATA
Cl. Holothuroidea

Dimensions
Up to 25 cm (10 in) long.
Distribution
Northeastern Pacific from Alaska to central California.
Food
In nature: plankton. In aquarium: juice from squashed bivalves, *Artemia*, chopped seafood. When a tentacle is saturated with food, it is drawn into the mouth and cleaned.
Notes
This species needs cold water, at temperatures of below 10°C (50°F).

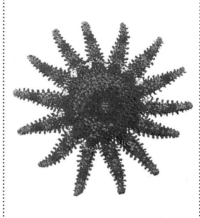
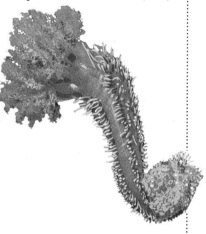

CYPRAEA sp.
Cl. Gastropoda

Dimensions
The largest species are up to 13 cm (5¼ in) long.
Distribution
Warm and tropical seas.
Food
In nature: algae and detritus. In aquarium: as in nature; also behaves as a cleaner.
Notes
While the majority of *Cypraea* species are found in tropical seas, some are also found in the Mediterranean.

DIADEMA ANTILLARUM
Cl. Echinoidea

Dimensions
Spines up to 30 cm (12 in) long.
Distribution
West Indies and Florida Keys.
Food
In nature: algae. In aquarium: algae and chopped seafood.
Notes
The spines are extremely sharp and fragile. They contain poison which, although not fatal, may cause a painful wound.

EUCIDARIS TRIBULOIDES
Cl. Echinoidea

Dimensions
Up to 6 cm (2½ in) in diameter.
Distribution
From South Carolina to Brazil.
Food
In nature: omnivorous, feeding on plant and animal substances. In aquarium: algae and chopped seafood.
Notes
This is one of the tropical reef sea urchins available and of interest to the marine aquarist.

HERMISSENDA CRASSICORNIS
Cl. Gastropoda

Dimensions
Up to 5 cm (2 in).
Distribution
North Pacific; Puget Sound.
Food
In nature: eggs, hydroids, ascidians, molluscs and similar soft-bodied marine invertebrates; may behave as a cleaner. In aquarium: small items of seafood, whole or in pieces.
Notes
The majority of nudibranch species are difficult to keep in captivity because of their food habits, but *Hermissenda* is less demanding and lives successfully in a cold-water aquarium at about 10°C (50°F). Individuals caught in shallow water south of Point Conception (California) tolerate warmer water.

HYMENOCERA PICTA
Cl. Crustacea

Dimensions
Up to 6 cm (2½ in).
Distribution
Tropical Indo-Pacific region.
Food
In nature: feeds exclusively on starfish. In aquarium: starfish and various types of seafood.
Notes
It will live happily with other tranquil, small-mouthed species, in an aquarium containing plenty of rocks and corals for shelter. The ideal water temperature is 24° – 28°C (75° – 82°F).

LINCKIA sp.
Cl. Asteroidea

Dimensions
Up to 20 cm (8 in).
Distribution
Tropical seas.
Food
In nature: omnivorous. In aquarium: chopped seafood.
Notes
The majority of *Linckia* species are sturdy and live well at temperatures of around 24°C (75°F). They are easy to keep in the aquarium.

LYSMATA WURDEMANNI
Cl. Crustacea

Dimensions
Up to 7 cm (2¾ in).
Distribution
From Chesapeake Bay to Brazil.
Food
Ectoparasites, organic detritus, small encrustant organisms.
Notes
Of the other species belonging to the genus, *Lysmata grahami*, from the West Indies, is particularly colourful and attractive.

MUREX sp.
Cl. Gastropoda

Dimensions
Some species up to 13 cm (5¼ in).
Distribution
Tropical and cold seas all over the world.
Food
In nature: preys on other molluscs, fish and small marine animals. In aquarium: live or chopped seafood.
Notes
The shells are very ornamental and thus greatly appreciated by collectors. *Murex* needs good quality water; the temperature depends on the species and original natural surroundings.

OCTOPUS BIMACULATUS
Cl. Cephalopoda

Dimensions
Up to 75 cm (30 in) long.
Distribution
From southern California to Mexico.
Food
In nature: crabs, shrimp, fish and molluscs. In aquarium: prefers live food.
Notes
This species has large spots on the body, normally ringed in blue with a darker center. Temperatures suitable for tropical aquaria are recommended.

PAGURUS ACADIANUS
Cl. Crustacea

Dimensions
Up to 5 cm (2 in).
Distribution
From Florida to Texas, including Bahamas and West Indies.
Food
In nature: cleaner, or predator of fish, worms and other marine organisms. In aquarium: chopped seafood.
Notes
As a rule, hermit crabs are easy to keep, but as they grow they have to be furnished with increasingly larger shells which they can adopt as homes.

PATIRA MINIATA
Cl. Asteroidea

Dimensions
Up to 10 cm (4 in) in diameter.
Distribution
From Alaska to Lower California and Revillagigedo Islands (Mexico).
Food
In nature: omnivorous; algae, sponges, sea urchins, eggs. In aquarium: algae, pieces of seafood.
Notes
In captivity this is one of the most hardy starfish, adapting to room temperatures.

PETROLISTHES sp.
Cl. Crustacea

Dimensions
Normally up to 2.5 cm (1 in).
Distribution
Coral seas; some species in temperate zones.
Food
In nature: plankton and small organisms. In aquarium: finely chopped seafood.
Notes
They live in both cold and warm seas and are often associated with sea anemones on coral reefs.

RADIANTHUS sp.
Cl. Anthozoa

Dimensions
Up to 25 cm (10 in) and more in diameter.
Distribution
Coral reefs of the tropical Indo-Pacific region.
Food
In nature: fish and other coral reef organisms. In aquarium: whole or chopped seafood, placed near tentacles.
Notes
Many large sea anemones live in symbiosis with other animals, as is the case with *Radianthus* which is host to various species of *Amphiprion*.

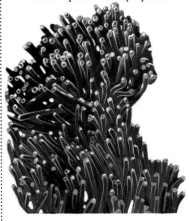

SPIROBRANCHUS GIGANTEUS
Cl. Polychaeta

Dimensions
Segments up to 2 cm (¾ in).
Distribution
Tropical coral reefs.
Food
In nature: plankton. In aquarium: *Artemia* nauplii and juice from squashed bivalves.
Notes
This species usually burrows among coral. Like all worms living in tubes, when they extend outside the tube, food should be placed close to the tentacles.

SPIROGRAPHIS SPALLANZANII
Cl. Polychaeta

Dimensions
Up to 30 cm (12 in) long.
Distribution
Mediterranean, European Atlantic coasts, North Sea.
Food
In nature: it feeds by filtering organic particles and minuscule planktonic organisms from the water. In aquarium: *Artemia* and the juice from chopped mussels.
Notes
If well fed, it adapts quite easily to aquarium life.

SQUILLA EMPUSA
Cl. Crustacea

Dimensions
Up to 25 cm (10 in).
Distribution
Atlantic coasts from the United States to Brazil.
Food
In nature: digs burrows to await prey; small fish and other organisms. In aquarium: seafood, live or dead, whole or in pieces.
Notes
Small varieties that live around reefs are available to aquarists but no matter what the size, care must be taken to avoid the strong claws which can cause painful wounds.

STENOPUS HISPIDUS
Cl. Crustacea

Dimensions
Up to 8 cm (3 in) long.
Distribution
Warm seas.
Food
In nature: a cleaner shrimp, ridding fish of ectoparasites. In aquarium: finely chopped seafood.
Notes
This shrimp sets up a cleaning station in the natural reef and in the aquarium; it removes parasites from fish which therefore pay it regular visits.

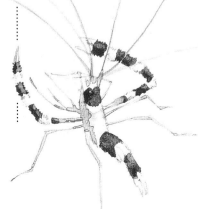

STOICHACTIS sp.
Cl. Anthozoa

Dimensions
Up to 1.5 cm (¾ in) in diameter.
Distribution
Tropical coral reefs.
Food
In nature: preys on fish and other reef organisms. In aquarium: whole live or chopped items of seafood, placed close to tentacles.
Notes
This is a giant sea anemone, fairly hardy if kept properly in the aquarium.

TEALIA LOFOTENSIS
Cl. Anthozoa

Dimensions
Up to 10 cm (4 in) in diameter.
Distribution
From state of Washington to San Diego, California.
Food
In nature: all kinds of marine organisms. In aquarium: chopped seafood.
Notes
Its area of distribution covers shallow intertidal waters to those 15 m (50 ft) deep. It lives successfully in the aquarium, preferring temperatures below 15°C (59°F).

TOXOPNEUSTES sp.
Cl. Echinoidea

Dimensions
Up to 13 cm (5¼ in) in diameter.
Distribution
Temperate and tropical oceans.
Food
In nature: omnivorous, feeding on algae and detritus. In aquarium: algae and pieces of seafood.
Notes
This group of sea urchins possesses poisonous pedicellariae which, in some species, are decidedly dangerous. They are not recommended for home aquaria.

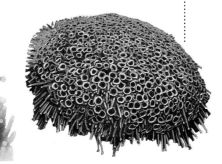

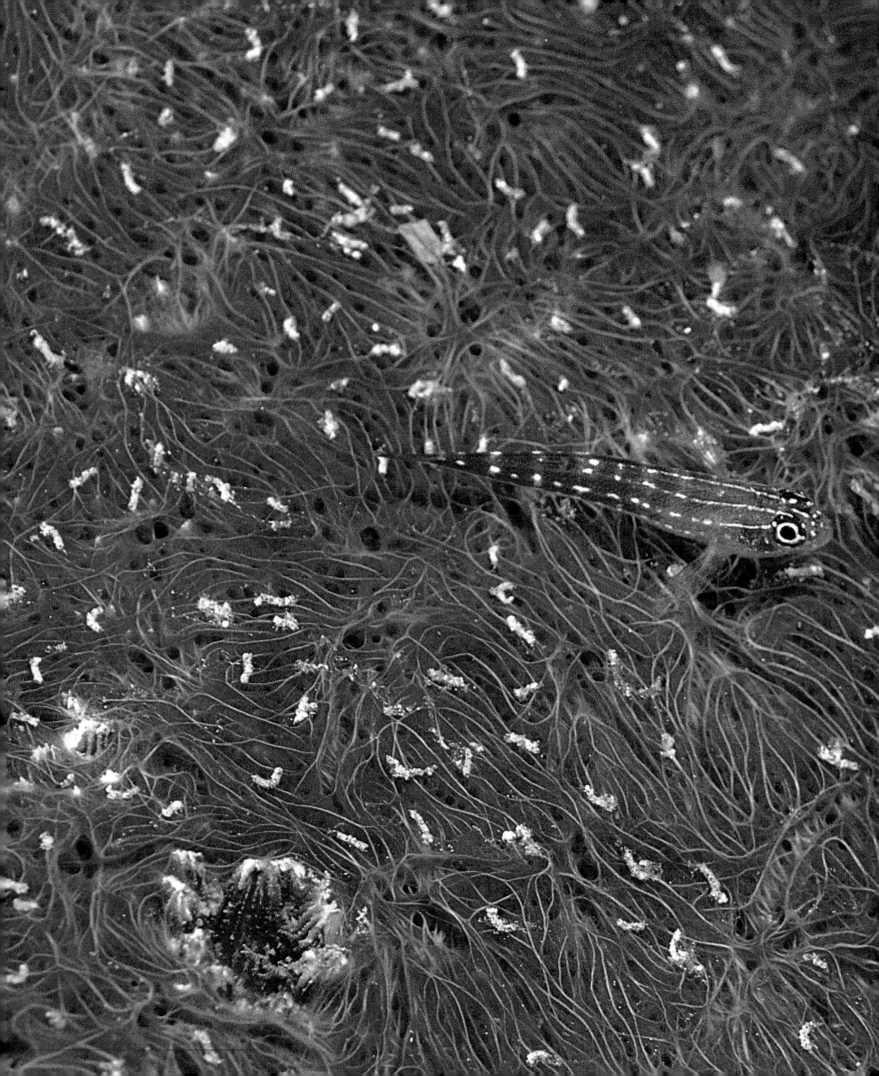

APPENDIX

Families
∎
Disease guide
∎
Glossary
∎
Index
∎
Bibliography

CLASSIFICATION OF THE SUPERCLASS OF FISH

Class	Subclasses	Orders			
CHONDRICHTHYES	PLAGIOSTOMATA AND HOLOCEPHALI	Squaliformes	Rajiformes	Chimaeriformes	
OSTEICHTHYES	BRACHYOPTERYGII	Polypteriformes			
OSTEICHTHYES	ACTINOPTERYGII	Lepisosteiformes or Ginglymodi	Amiiformes or Protospondyli	Acipenseriformes	Clupeiformes
		Myctophiformes	Esociformes	Mormyriformes	Cypriniformes
		Siluriformes	Anguilliformes	Cyprinodontiformes	Beloniformes
		Macruriformes	Gadiformes	Lampridiformes	Beryciformes
		Perciformes	Echeneidiformes	Zeiformes	Pleuronectiformes
		Gasterosteiformes	Syngnathiformes	Mugiliformes	Lophiiformes
		Batrachoidiformes	Gobiesociformes	Tetraodontiformes	
OSTEICHTHYES	CROSSOPTERYGII	Coelacanthiformes	Dipnoi or Ceratodiformes		

Families of tropical and temperate freshwater fish

Fish represent more than 50 per cent of all known vertebrates. Of the 22,000 or so recognized species, some 97 per cent belong to the Teleostei, those fish with a bony skeleton. Almost all aquarium fish are to be found in this category. The table opposite gives the classification of the superclass of fish and the orders into which it is divided.

Even though freshwater streams, rivers and lakes constitute a mere 1 per cent of the earth's surface, the fish that live in them are not significantly fewer in number than those living in the seas and oceans, representing, in fact, about 40 per cent of the total. So if one compares the number of species with the quantity of water available, their population appears to be excessive in relation to that of marine fish. Speciation – biological processes leading to the formation of new species – has therefore operated with greater efficiency in fresh water, particularly in the tropical regions where the number of fish species sometimes exhibits a diversity comparable to that found in and around a coral reef.

DIPNOI
Cl. Osteichthyes

Although seldom considered for a domestic aquarium, the Dipnoi, or lungfish, are of particular interest because they are of extremely ancient origin, going back some 300 million years. Their specialized lungs probably developed as adaptations to stagnant waters and drought, perhaps in the Devonian, when the water levels fell.
Today there are six known living species (one in Australia, four in Africa and one in South America), indigenous to the tropical regions subject to seasonal drought. The Australian dipnoid (*Neoceratodus*) is regarded as the most primitive, possessing a single lung. The African (*Protopterus*) and South American (*Lepidosiren*) Dipnoi possess two lungs; during the dry season they estivate in tunnels or balls of mud; in water they have to come to the surface to breathe, otherwise they die of asphyxiation. If kept in an aquarium, an air space must be left between the water surface and the cover. The Dipnoi are oviparous and lay eggs in the wet season. The eggs as well as the young are normally guarded by the male.
Neoceratodus forsteri, because of its rarity, is one of the species protected by the Convention on International Trade and Endangered Species (CITES) in respect of capture, commerce and confinement.
☐
Lepidosiren paradoxa, p. 51
Neoceratodus forsteri, p. 53
Protopterus dolloi, p. 57

OSTEOGLOSSIDAE
Ord. Osteoglossiformes

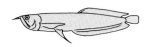

The Osteoglossidae make up a small family of freshwater fish found in tropical zones. Fish fossils belonging to this family have been discovered in Eocene deposits in Wyoming and Sumatra, the latter being where the Osteoglossidae of the genus *Scleropages* still live today. Most members of the family are of large dimensions, with predatory habits, as is evidenced by their big mouth, and a compressed, rounded body. The fish have projecting scales, large eyes, an armour-plated head, and dorsal and anal fins placed far back. The name of the family is derived from the strong bony structure that supports the tongue. One member of the family (although some students place it in the separate family Arapaimidae) is *Arapaima gigas*, which measures 4.5 m (15 ft) in length and is one of the biggest fish in the world. The family also contains the popular aquarium fish *Osteoglossum bicirrhosum*, as well as the related *O. ferreirai* and the two *Scleropages* species, found in northern Australia and Southeast Asia.
Both A. gigas and *Scleropages formosus* are protected by CITES in respect of capture, commerce and confinement.
☐
Osteoglossum bicirrhosum, p.54

NOTOPTERIDAE
Ord. Osteoglossiformes

The Notopteridae live in the tropical fresh waters of Africa and Southeast Asia. They are fairly large fish, elongated and laterally compressed. Because of their appearance, they are sometimes known as knife fish. Their oddest feature is the very long anal fin which extends from the head and along the whole abdomen. The dorsal fin is small and pinnate, except in one African species, where it is wholly missing; the ventral fins are small or absent, but the pectoral fins are large. The anal fin is the main organ of propulsion, and the fish can swim both forward and backward. There are three known genera with four species that are relatively big, ranging in length from 20 cm (8 in) to 90 cm (36 in). Only the young are adapted to living in domestic aquaria.
☐
Notopterus chitala, p. 54

PANTODONTIDAE
Ord. Osteoglossiformes

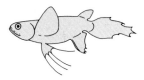

This family, present only in the waters of West Africa, comprises only one species, *Pantodon buchholzi*, commonly known as the butterfly fish. The ventral fins of the fish consist of rays linked by the interradial membrane only at the base, so that they extend downward, independent and digitiform, for most of their considerable length. The pectoral fins, however, are very large and broad, like butterfly wings, and used by the fish to leap from the water to a height of some 2 m (6½ ft). The big mouth, directed upward to facilitate the capture of prey, contains a very large number of teeth which also cover the tongue. During the day butterfly fish normally conceal themselves under floating leaves and detritus, feeding mainly in the evening and at night. They consume surface-dwelling animals, including terrestrial insects that fall into the water accidentally. The species displays characteristic sexual dimorphism, the male being distinguished from the female by his concave anal fin.
☐
Pantodon buchholzi, p. 55

MORMYRIDAE
Ord. Osteoglossiformes

This family, commonly known as elephant fish, is found in tropical fresh waters of Africa, and it includes fish that can generate electric shocks, used to locate prey and objects situated in the vicinity.
Many of these fish have a unique appearance, with small eyes and often a strange mouth with elongated jaws that form a proboscis. The dorsal and anal fins are usually placed far back, so much so that the body narrows sharply towards the tail, which is markedly bifurcate. The mouth contains a few small teeth. The Mormyridae have an uncommonly large cerebellum, a swim bladder that extends to the skull, and tail muscles that are modified into an electric organ. They normally live in murky swamps, along the banks of slow-flowing rivers and in fresh waters with a muddy bottom. They are mainly nocturnal, but those species which live in dark surroundings are also active by day. Their food consists principally of worms and other sedentary invertebrates living on surface mud. Little is known of their reproductive behaviour, but apparently at least a few species build bubble nests. There are some 16 known genera and about 190 species.
□
Gnathonemus petersi, p.48

CHARACIDAE
Ord. Characiformes

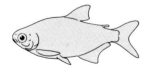

The Characidae constitute one of the largest families of fish, comprising many hundreds of species. They are an important group for the aquarist because many species are small, colourful and active, cheap to buy and easy to raise. With few exceptions, such as the piranha, the characins are tranquil and suitable either for being kept on their own or together with other species. Classification of this group is complicated and particularly confusing for non-specialists. The Characiformes, along with other orders including the orders Cypriniformes and Siluriformes, make up the superorder Ostariophysi. These fish are characterized by Weber's apparatus (constituted of a series of tiny bones which connect the swim bladder with the inner ear, through which sounds are enhanced) and by the secretion of an alarm substance, emitted when the skin is damaged, which warns other members of the species.
The Characidae may be fusiform (neon tetra) or compressed (piranha); they almost always have scales, usually cycloid, rarely ctenoid, and large teeth, but no barbels. As a rule they possess an adipose fin, ventral fins in an abdominal position, and a short to medium anal fin. They are commonly found in the tropical fresh waters of Africa and South America. Most of them are carnivorous. They reproduce by laying eggs on bottom vegetation and in open waters.
□
Alestes longipinnis, p. 38
Anoptichthys jordani, p. 39
Astyanax mexicanus, p. 40
Chalceus macrolepidotus, p. 44
Cheirodon axelrodi, p. 44
Colossoma brachypomum, p. 45
Exodon paradoxus, p. 47
Gymnocorymbus ternetzi, p. 48
Hemigrammus ocellifer, p. 49
Hyphessobrycon pulchripinnis, p. 49
Hyphessobrycon serpae, p. 50
Metynnis hypsauchen, p. 52
Paracheirodon innesi, p. 55
Phenacogrammus interruptus, p.56
Serrasalmus nattereri, p. 59

LEBIASINIDAE
Ord. Characiformes

The Lebiasinidae, once included in the family Characidae, are today classified separately in their own right. These fish have a terminal mouth which in the majority of species does not extend to the eye; a short dorsal fin and small pectoral fins; and sometimes an adipose fin and an incomplete or absent lateral line, as in the case of *Pyrrhulina*. The family comprises the genera *Copeina*, *Copella*, *Pyrrhulina* and *Nannostomus*. They inhabit the northern part of South America, including Guyana, Suriname and the Amazon basin. Most species are adapted to living near the water surface, where they find the organisms on which they feed. Reproduction is typical of the Characidae, with short periods of development and, in some instances, care of eggs.
□
Copeina arnoldi, p. 46
Nannostomus harrisoni, p. 53

ANOSTOMIDAE
Ord. Characiformes

This family, similar in many ways to the Lebiasinidae, comprises two of the best known and most popular aquarium fish, *Leporinus* and *Anostomus*.
The Anostomidae are usually elongated, with a short anal fin, tubular nostrils, a small, nonprotrusile mouth, a straight lateral line and normally just three or four incisor teeth in either jaw. Although the genus *Abramites* has a relatively high body, it also belongs to this family.
The Anostomidae have a wide geographical range which extends from Central America to the West Indies and to Argentina. Generally they are gregarious, preferring enclosed or slow-running forest waters. Many species avoid strong light. The fish are normally omnivorous, especially in the wild, where they feed on worms, larvae, algae, etc. The body is shaped in such a way that they can dart from cover with rapid acceleration, when they feel threatened. Many species, even small ones, are hardy, adapting to considerable variations in water conditions, temperature and number of aquarium companions. Some species are liable to be aggressive. Little is known, in most cases, of reproductive behaviour. *Leporinus fasciatus*, however, is bred commercially.
□
Anostomus anostomus, p. 39
Leporinus fasciatus, p. 51

GASTEROPELECIDAE
Ord. Characiformes

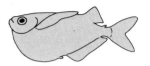

The freshwater Gasteropelecidae are a small family of fish distinguished by their straight back, highly compressed body and large pectoral fins attached to a muscular and wedge-shaped chest. They are considered the only fish that are literally able to fly, rather than glide. Thanks to the well-developed muscles, they flap their pectoral fins and in this way actually make flights of 15 m (50 ft) or more close to the water surface. The lower lobe of the caudal fin remains in contact with the water during part of, or the whole of, the flight. The family comprises three genera: *Carnegiella*, *Gasteropelecus* and *Thoracocharax*, with nine species in all.
The fish are found in tropical America from Panama to Argentina, including Panama's Rio Bayano, the rivers of the Orinoco and Amazon basins, the Rio Paraguay and the Rio de la Plata. They are normally present in slow-flowing streams and rivers, stagnant water, swamps and lakes. Gregarious by habit, they feed on insects at the surface. It is not known whether they take flight to escape predators or to catch food.
□
Carnegiella strigata, p. 44

CYPRINIDAE
Ord. Cypriniformes

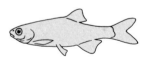

This is the largest family of fish, comprising more than 2,000 species. They vary in size from about 4 cm (1½ in) to 2.5 m (8 ft). The Cyprinidae are as a rule moderately compressed, with a protrusile mouth, lacking teeth in the jaw or on the palate, but with three rows in the throat (pharyngeal teeth). The fins do not have true spines (unsegmented) although there may be one or two stiff rays (segmented); there is no adipose fin. The tail is normally bifurcate. The scales are cycloid and do not cover the head. The typical aquarium cyprinid has large, shiny scales, a single dorsal fin halfway down the body, and a small terminal mouth. The family belongs to the order Cypriniformes which no longer includes the Characidae, now given their own order, Characiformes. Both orders remain in the superorder Ostariophysi, characterized by the presence of Weber's apparatus and an alarm substance secreted when the skin is damaged.

The Cyprinidae are plentiful in tropical and temperate waters throughout much of the world, but are not native to South America, Australia and Madagascar. As might be expected in a family of this size, its members are found in any type of freshwater habitat where fish life is possible: streams, rivers, lakes, ponds and pools, and at temperatures varying from around 0°C (32°F) to 38°C (100°F). They reproduce by laying eggs among the plants, and many males develop "reproductive" tubercles when they reach sexual maturity.

☐

Barbus conchonius, p. 41
Barbus cumingi, p. 41
Barbus nigrofasciatus, p. 41
Barbus pentazona, p. 41
Barbus schwanenfeldi, p. 42
Barbus tetrazona, p. 42
Brachydanio albolineatus, p. 43
Brachydanio rerio, p. 43
Labeo erythrurus, p. 51
Rasbora heteromorpha, p. 57
Rasbora maculata, p. 58
Rasbora trilineata, p. 58
Carassius auratus, p. 62
Cyprinus carpio, p. 62
Rhodeus sericeus, p. 63

COBITIDAE
Ord. Cypriniformes

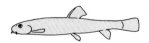

In spite of their worm-like body, the Cobitidae, or loaches, are closely related to the Cyprinidae, sharing certain characteristics such as the Weber apparatus, the alarm substances and the pelvic fins in an abdominal position. They possess small eyes, tiny scales that may appear invisible to the naked eye, and a small, subterminal mouth without teeth. There are three, four or more pairs of barbels, of which at least one is on the snout, and a protective erectile spine under the eye. The small swim bladder does not permit the fish to float and constitutes a typical adaptation of fish that live on the bottom. Furthermore, it seems that the swim bladder makes some Cobitidae sensitive to atmospheric pressure. For example, *Misgurnus anguillicaudatus* becomes more active when the air pressure falls. There are at least 175 species of Cobitidae, widely distributed in Europe and Asia; they are absent in America. Those species of worm-like appearance also have a flattened belly.

☐

Acanthophthalmus semicinctus, p. 38
Botia macracantha, p. 42

SILURIDAE
Ord. Siluriformes

The catfish of the family Siluridae belong to the order Siluriformes together with about 30 other families of catfish. They, like the Characiformes and Cypriniformes, are furnished with Weber's apparatus, and if wounded, emit an alarm substance. They are also part of the superorder Ostariophyti. Catfish lack scales but may be covered with bony plates; normally there is an adipose fin and often spines in front of the pectoral and dorsal fins. As a rule they have barbels on the jaws, and small eyes. Some species are poisonous, containing toxic cells in the epidermis which covers the fin spines. Catfish may be quite large – *Silurus glanis* from Europe measures at least 3.3 m (11 ft) in length – but most are barely 12 cm (5 in). The Siluridae are confined to the temperate and tropical waters of Europe and Asia. Only a few species are of interest to aquarists.

☐

Kryptopterus bicirrhis, p. 50

CLARIIDAE
Ord. Siluriformes

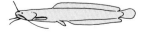

The catfish of the family Clariidae have a very long dorsal fin, a rounded caudal fin and four pairs of barbels around the mouth. One of the principal features of the family is the presence of supplementary labrinthine breathing organs. These are situated in the cavities that communicate with the gill chambers and they permit the fish to remain out of water and to survive for some time in poorly oxygenated water. The structure enables the various species of the genus *Clarias* to emerge from the water and move by land from one watery environment to another, thus colonizing extensive territories. Not all the Clariidae, however, possess these organs. They are not found, for example, in those species that live in turbulent, well-oxygenated water. The pectoral fins of the fish are provided with a spiny ray linked to poison glands which can inject a secretion that may cause severe pain and symptoms of paralysis in the affected part. The Clariidae are present in Africa, the Middle East and Asia. The 100 known species are divided into 13 genera.

☐

Clarias batrachus, p. 45

MOCHOKIDAE
Ord. Siluriformes

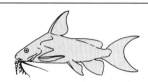

The members of this family of catfish have a strange adipose fin containing bony rays and an anal fin with no fewer than nine rays. They are found only in Africa. As is normal, they have no scales. The dorsal and pectoral fins are armed with spiny, toothed rays, linked to poison glands. Should these fish be threatened, rays can be locked into an erect position so as to form the vertices of a triangle and increasing the body dimensions. The mouth is surrounded by three barbels sensitive to touch and taste; one long pair above and two shorter pinnate pairs below. Numerous members of the family, which includes more than 80 species belonging to the genus *Synodontis*, are known as "squeakers" because of the trilling sounds they emit, due, it is thought, to the rotation of the spiny rays in their alveoli. In nature these fish dislike the light and are most active at dusk. During the day they stay hidden in the shade of a hole, branches or large submerged leaves, adhering to them by their belly, maintaining an upside-down position. They often congregate in large shoals and live in streams and rivers with weak currents. Altogether, more than 150 species are known.

☐

Synodontis alberti, p. 59
Synodontis nigriventris, p. 59

MALAPTERURIDAE
Ord. Siluriformes

This small family contains only two species, *Malapterurus electricus* and *M. microstoma*, both African, known as electric catfish. They are thick and oblong, without scales and dorsal fin. The pectoral fins lack spines, the adipose fin is set well back, and the thick, fleshy lips bear three pairs of barbels. The electricity is generated by a subcutaneous organ which covers much of the body's muscle system. Polarity is negative at the front and positive in the tail region. The shock is apparently delivered by a large nerve cell which originates in the spinal medulla and acts as an interrupter. The shocks, of up to 350 volts, are used to stun prey and discourage predators. These fish are not normally kept in home aquaria because the adults reach a considerable size, but they are very interesting for research work into special adaptations. In public aquaria they are sometimes exhibited so as to highlight their electrical properties. When offered their food, they are accustomed to giving out a shock which can be recorded by means of electrodes, placed in the aquarium, and then transmitted to oscilloscopes, loudspeakers and other forms of apparatus capable of measuring and relaying them.

☐

Malapterurus electricus, p. 52

CALLICHTHYIDAE
Ord. Siluriformes

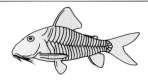

The principal feature of this family, the mailed catfish, is the presence of two rows of bony plates on the flanks, arranged like the tiles of a roof. Sometimes these plates are also found on the head and back. The large dorsal fin and the adipose fin are characterized by strong mobile spines. There are two to four pairs of barbels (the number varies according to the genus concerned) on either side of the mouth. The family comprises 110 species (genera *Callichthys*, *Corydoras*, *Hoplosternum*, etc.), some of which make excellent aquarium subjects. The Callichthyidae originate in South America and Trinidad. The fish live in groups, normally at no great depth, in gently flowing streams and rivers with muddy or sandy bottoms, or, more rarely, in ponds. A certain number of species have been raised successfully in captivity; some build bubble nests. Several members of the family are capable of moving from pond to pond over wet ground by means of their pectoral spines and wriggling movements of the body. The species of the genus *Corydoras*, in particular, can spend long periods out of water, breathing air from the atmosphere; in fact, they are capable of swallowing air and absorbing the oxygen contained therein through the vascular walls of their stomach and intestine.
□
Callichthys callichthys, p. 43
Corydoras aeneus, p. 46

LORICARIIDAE
Ord. Siluriformes

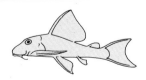

The catfish of the family Loricariidae are similar in all respects to those of the Callichthyidae, but they lack the conspicuous barbels around the sucker-like mouth; they have three or four rows of bony plates on either flank. There are at least 450 species, the majority of which live in the fresh waters of South America, Panama and Costa Rica, while some are to be found in brackish water. The fish normally inhabit streams and rivers, often torrential, with rocky or stony bottoms, feeding principally on algae, which they scrape from the substrate with their ventrally positioned mouth. Some claim that the strong bony plates serve to protect the fish from the abrasive action of the sediment swept along by the current. Active particularly at dusk, they possess a membrane which they can lower over the pupil, thus reducing the amount of incident light when this becomes too intense. Some species of Loricariidae are capable of giving out sounds that sometimes recall cries emitted by other animals.
□
Hypostomus plecostomus, p. 50
Loricaria filamentosa, p. 52

GYMNOTIDAE
Ord. Gymnotiformes

This small family of eel-like fish is confined to the fresh waters of Central and South America, having probably evolved from characoid fish. Only three species are known, all belonging to the genus *Gymnotus*. They have a compressed, eel-shaped body covered with small scales, and they move by waving their long anal fin in such a way as to propel themselves either forward or backward. They lack dorsal, caudal and pelvic fins. The Gymnotidae possess weak electric organs which help them to find their location in dark waters and at night. The electric shocks vary from 5 to 190 per second when the animals are calm and become more frequent when excited. Laboratory experiments show that the electric field so produced is also used for finding food and locating predators. The Gymnotidae are in fact hunted by many fish and possibly for this reason they are able to regenerate any parts of their body, especially the tail. They also possess supplementary breathing organs which enable them to gulp air and thus to survive in poorly oxygenated water.
□
Gymnotus carapo, p. 48

CYPRINODONTIDAE
Ord. Cyprinodontiformes

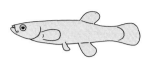

The family Cyprinodontidae comprises approximately 300 species, many of which are of interest to aquarists. They are small, seldom measuring over 13 cm (5¼ in), and often brightly coloured. They are similar to the Cyprinidae, differing only in that they possess teeth in both jaws. The head is often flattened, with a terminal or superior mouth, without barbels. The single dorsal fin is set far back on the body, there is no adipose fin, the pelvic fins may or may not be present, and the tail is normally rounded. The lateral line is limited mainly to the head. The Cyprinodontidae are found in a variety of habitats, from desert springs and temporary tropical pools to high mountains, lakes, marshes and coastal swamps. Of particular interest to aquarists are those African and South American species that live annually or seasonally in temporary pools. These fish are annual in the literal sense of the word; they lay their eggs in the substrate and die when the pond dries up. The eggs survive until the next rainy season, when they hatch and repopulate the pool. Although the Cyprinodontidae lay eggs, fertilization is internal. Depending upon the species, the eggs are laid either among vegetation or in the substrate. Some follow a normal aquatic incubation process, while others need a period of substrate drought in order to develop. The intermediate forms have eggs that tolerate, but do not depend upon, a dry period.
□
Aphyosemion australe, p. 39
Aphyosemion sjoestedi, p. 40
Cyprinodon macularis, p. 46
Fundulus chrysotus, p. 47
Oryzias latipes, p. 54

POECILIDAE
Ord. Cyprinodontiformes

These small fish, rarely more than 10 cm (4 in) long, are very similar in looks to the Cyprinodontidae. The body is stocky, as a rule slightly compressed, covered with large cycloid scales. There are no fin spines and the single dorsal fin is positioned halfway down the body, to the rear. The tail is usually rounded. The mouth is terminal, generally opening upward, with teeth in both jaws. The male's anal fin is modified to form a copulatory organ, or gonopodium, a kind of funnel through which the packets of sperm, or spermatophores, are inserted in the female. The sperm may be stored, and as a result many clutches of eggs may be fertilized by the single act of copulation.
The Poecilidae are found in fresh and brackish waters of the western United States as far as Argentina, but only a few species live south of the Amazon basin. They are very common in lakes and swamps and along river banks. Some species, such as *Gambusia affinis*, are very important in the fight against mosquitos. Because of sexual dimorphism, artificial selection is a simple procedure and many domestic varieties have been produced. The females are often cannibals and should be removed from the aquarium after the birth of the fry or kept in a trap with a slit in the floor through which the young can escape.
□
Poecilia reticulata, p. 56
Poecilia sphenops, p. 56
Poecilia velifera, p. 56
Xiphophorus helleri, p. 61
Xiphophorus maculatus, p. 61
Gambusia affinis, p. 62

ANABLEPIDAE
Ord. Cypriniformes

The Anablepidae are closely related to the Cyprinodontidae and Poecilidae. They are surface-dwelling fish with eyes adapted for vision both in air and water. Each eye has two pupils and a divided cornea. In the fish's normal position, one pupil is above and one below the water surface. Light rays coming from above pass through the upper pupil and are focused on the lower retina through the small extremity of a single pear-shaped lens; the rays coming from below the surface, however, pass through the larger and more curved diameter of the lens to focus on the upper retina.
There are three species, all

inhabiting the central and northern parts of South America, where they are found in lakes, rivers, canals and ditches, coastal lagoons, estuaries and, in the case of *Anableps microlepis*, in the sea just offshore. The water can be either cloudy or clear. During the day they move about in groups, feeding on surface organisms; by night they seek deeper waters, among rocks and plants, feeding on insects and crustaceans. The fish have remarkable jumping powers, using this ability as well as their excellent vision to escape predators, including birds. Like the Poecilidae, they are ovoviviparous with internal fertilization. The male has a well-developed gonopodium which can only move to one side; the female has unpaired genital openings. It is often claimed that males with right-facing genital organs can copulate only with "left-handed" females and vice versa, but this has not been verified.
☐
Anableps anableps, p. 39

GOODEIDAE
Ord. Cyprinodontiformes

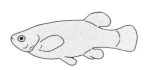

The Goodeidae are separated from their close relatives, the Cyprinodontidae and Poecilidae, as a result of their peculiar reproductive habits. As in the latter family, fertilization is internal and fully formed larvae are born. However, in the Goodeidae the male does not have a large tubiform gonopodium but an intromittent organ made up of the first rays of the anal fin, which are stiff and separated by a notch. The embryos, too, are strange inasmuch as all species, except one, have nastriform extensions of the rectum, called trophotaeniae, which enable them to absorb food and oxygen from the female and to eliminate waste matter.
The fish have a terminal mouth with an elongated lower jaw and teeth on either sides of the jaws. The single dorsal fin is situated to the rear, as is the anal fin. There are pelvic fins but no adipose fin and fin spines. The scales are large and cycloid. Most species measure from 3.5 cm (1½ in) to about 10 cm (4 in), with a stocky or slim body.
There are about 35 species, confined to the highlands of central Mexico. At least 20 species have been raised successfully by aquarists. They live in various habitats, ranging from rocky mountain streams with little vegetation to village-polluted rivers with muddy bottoms and plenty of organic material, including algae. After fertilization, the incubation period lasts about 45 days, but this may fluctuate between 4 and 11 weeks, according to species and temperature. The number of eggs varies from fewer than 10 to more than 40. The young survive on good-quality food for fry.
☐
Ilyodon furcidens, p. 50

COTTIDAE
Ord. Scorpaeniformes

The freshwater Cottidae are small bottom-dwelling fish. The big flattened head, the large pectoral fins and the absence of a swim bladder help them to adapt to life on the bottom and enable them to retain their stability in fast-flowing, turbulent waters. The mouth is large, normally terminal, and the eyes are dorsally situated. The scales are ctenoid, generally in several rows, but sometimes they are reduced to spines or totally lacking. The spiny parts and soft rays of the dorsal fin are divided.
The Cottidae are oviparous and fertilization is internal; the males have a conspicuous genital papilla. Most of them are, however, marine fish, and the freshwater species normally tolerate a high degree of salinity. Some, in fact, spend time in both environments. There are some 65 known genera and about 300 species. Basically they are cold-water fish, plentiful in North America, Europe and Asia. In New Zealand, Australia and Argentina, only one species is found.
☐
Cottus gobio, p. 62

GASTEROSTEIDAE
Ord. Gasterosteiformes

The sticklebacks are known chiefly for their nest-building and reproductive behaviour. There are five genera, but only eight species. They are small fish with a long, tapering body, laterally compressed, with a terminal mouth opening upwards. They are covered with a series of bony plates rather than scales, and further protected by a line of spines along the back as well as pelvic fins which take the form of lateral spines. There are marine, freshwater and anadromous species, living as a rule in calm surroundings among thick vegetation. They inhabit North America, Europe and northern Asia.
The breeding behaviour of sticklebacks has been closely observed and widely studied. Having reached sexual maturity, the male takes on his spawning colours, characterized by a bright red belly, and builds a tunnel-shaped nest from vegetation, held together by his own sticky secretions. The gravid female is courted by the male and eventually persuaded, with nudges, to enter the nest. She lays her eggs, which are fertilized by the male, who normally repeats the operation with several females. He then defends the nest vigorously, fanning the eggs for six to eight days until they hatch. He will continue to oversee the fry until they are strong enough to fend for themselves.
☐
Gasterosteus aculeatus, p. 63

PERCIDAE
Ord. Perciformes

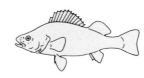

This family represents the prototype of an immense order of fish. They are predominantly marine forms, belonging to a number of families, whose characteristics include the presence of two dorsal fins, often close together and sometimes joined, the first being composed of spiny rays; one or two rays of the anal fin are likewise spiny. The ventral fins are situated below the pectorals, which are thus thoracic, and the scales are ctenoid. The swim bladder does not communicate with the oesophagus. They are predatory fish with a large, deeply notched mouth provided with small conical teeth. The Percidae are freshwater fish of the Northern Hemisphere. Palaeontological tests suggest that they originated as forms of Serranidae capable of living temporarily in fresh water, perhaps like present-day salmon. They probably appeared first in Europe, later ranging across northern Asia to North America over the land bridges which connected the two continents about 60 million years ago.
☐
Perca fluviatilis, p. 63

CENTROPOMIDAE
Ord. Perciformes

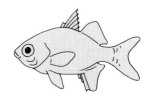

The Centropomidae comprise two groups of fish, apparently different yet with a number of features in common, such as the lateral line which extends to the tail. They have dorsal fins divided or joined by a small notch. These fish are found off the coasts of the tropical Atlantic and along the eastern shores of the tropical Pacific, where they enter estuaries and rivers. The family also contains the Nile perch, *Lates niloticus*, which may weigh almost 100 kg (220 lb). This huge species is normally kept in public aquaria. The much smaller, transparent members of the genus *Chanda* and *Gymnochanda* are suitable subjects for home aquaria.
☐
Chanda wolfii, p. 44

LOBOTIDAE
Ord. Perciformes

The Lobotidae are recognizable by their large, rounded dorsal and anal fins which, together with the round caudal fin, give the impression that the fish have three tails. The body is high and strongly compressed. Both jaws carry villiform teeth, the outer ones being bigger, divided into bands. The family comprises a few marine, brackish water and freshwater forms which are principally tropical. It includes the genera *Lobotes* with a single marine species (*L. surinamensis*), also found in the Mediterranean, and *Datnioides*, confined to the fresh and slightly brackish waters of the Indo-Pacific region, from Thailand to Australia.
☐
Datnioides microlepis, p. 46

CENTRARCHIDAE
Ord. Perciformes

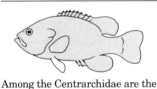

Among the Centrarchidae are the sunfish and the black basses, many of which are important quarry of sporting fishermen. Relatively large, they are generally kept in ponds by aquarists. Indigenous only to North America, there are nine genera and some 30 species, some of which have been introduced to Europe and become thoroughly acclimated.
The fish have a long or high body, slightly or markedly compressed. The undivided dorsal fin comprises both spines and rays. The rear part with soft rays is symmetrical with the large anal fin. The mouth is terminal, the upper jaw is protrusile, and the teeth are small. The scales are ctenoid and the lateral line is complete. Some species are notable for their particularly bright coloration, which becomes even more vivid in the spawning season. They are opportunistic carnivores, normally living in calm, slow waters; they build nests by digging small trenches on the sandy or gravelly bottom, and the females lay adhesive eggs. As a rule, the males guard the eggs until they hatch and then defend the fry.
☐
Lepomis gibbosus, p. 63

TOXOTIDAE
Ord. Perciformes

The archer fish owe their name to the characteristic shape of their mouth cavity, in which the tongue and the palate can form a kind of blowtube, enabling them to spray a jet of water strongly and accurately enough to stun an insect up to several feet above the surface. The species are of medium size, with a highly compressed body and an almost straight back. The mouth is large, terminal and freely protrusile. The big eyes are situated far forward. The dorsal fin is continuous, the pectoral fins are strong and well developed, the lateral line is complete and the scales are ctenoid.
The surface-dwelling archer fish are from the coastal waters of the Indo-Australian regions. The young are found both in fresh water and in estuaries. The adults are principally marine fish and lay their eggs on the coral reefs. All six known species belong to the single genus *Toxotes*.
☐
Toxotes jaculator, p. 60

MONODACTYLIDAE
Ord. Perciformes

This small family contains many attractive aquarium species, known as finger fish. They are coastal fish, living both in sea and brackish water, and sometimes, especially the young, in fresh water. They are in some ways similar to the Pomacanthidae and the Cichlidae of the genus *Pterophyllum* but have shorter dorsal and anal fins and a higher body, especially *Monodactylus sebae,* whose height exceeds its length. The head and mouth are comparatively small, and the dorsal and anal fins are partially covered with scales. The anterior spines of the dorsal fin are short and separated. The fish are as a rule silvery with broken vertical stripes. They are voracious and catch all sorts of small animals. They live in shoals in the more open and quiet waters at the mouths of rivers, although they normally prefer brackish zones. There are some five species, in three genera, inhabiting the coastal waters of Africa, except the Mediterranean, and as far as Southeast Asia and tropical Australia.
☐
Monodactylus argenteus, p. 53

SCATOPHAGIDAE
Ord. Perciformes

The scats make up a small family of fish that are similar in appearance to the Chaetodontidae, differing from the latter in that the mouth is not protrusile and the dorsal fin has a deep notch, with the spiny parts and soft rays almost divided. The two families are also alike in possessing ctenoid scales which are similarly distributed, to some extent also covering the dorsal and anal fins. There are no noticeable sexual differences. Originally from the tropical waters of Southeast Asia, the fish are found in marine, brackish and fresh water, but apparently only reproduce in the sea. Young scats go through an initial stage of development known as *Tholichthys,* once thought to be a separate species. During this early phase the head and neck are protected by a bony casing which disappears when adult. Spawning behaviour seems to resemble that of the Cichlidae, for it includes cleaning of the substrate and care of the young by both parents.
The family name means "eater of excrement" and refers to the fish's habit of consuming animal waste. As might be expected, therefore, the scats are abundant near drains and sewers. Despite their unhealthy habits in the wild, the fish are nevertheless attractive aquarium subjects.
☐
Scatophagus argus, p. 58

NANDIDAE
Ord. Perciformes

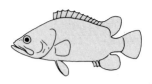

The Nandidae are tall, compressed fish, with a notable capacity for concealment and mimicry. The head and mouth are large and the jaws are highly protrusile. The dorsal fin is continuous and contains spiny and soft rays. The caudal fin is rounded. The lateral line is incomplete or absent. There are no external sexual differences. The fish live in shallow water where, thanks to their mimetic ability, they hide among reeds, branches and submerged roots. Whilst awaiting prey, they may swim slowly on one side, so they resemble a floating or sinking leaf, and then pounce upon any unwary fish. The ten or so known species are divided into seven genera, found in South America, West Africa and southern Asia. Various species are available to aquarists.
☐
Monocirrhus polyacanthus, p. 53

CICHLIDAE
Ord. Perciformes

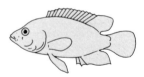

The Cichlidae are among the best known and most popular freshwater aquarium fish. They are sturdy, with a high, compressed body and conspicuous large or medium scales, normally ctenoid, that only partially cover the head. The lateral line is generally divided into separate parts, anterior–superior and posterior–inferior. The dorsal and anal fins have anterior spiny rays followed by soft rays. The soft rays of the dorsal, anal and pelvic fins may be elongate; the caudal fin is normally round or truncate, and there is no adipose fin. The teeth are often conical but vary considerably according to feeding habits. The mouth is protrusile. Like the Pomacentridae, they have a single nostril which opens on either side of the snout.
There are at least 1,000 species of Cichlidae throughout the world. They are indigenous to Africa, where most

of them are to be found, and they also range from South America to Texas, where one species exists. They are absent from most of Asia, and from Europe and Australia.

The majority of cichlids are predators, hunting mainly small fish and insects but they have evolved into many types, enjoying a variety of diets. In the aquarium they feed voraciously.

The Cichlidae are oviparous, laying eggs, and are generally noted for their territorial instincts and parental care. Various patterns of egg laying and subsequent raising of young have been developed:
1. Eggs laid on substrate; surveillance of eggs and young.
2. Eggs laid on substrate; surveillance of eggs; female mouth-broods the young.
3. Eggs and young mouth-brooded by both parents.
4. Eggs and young mouth-brooded by female alone.
5. Eggs and young mouth-brooded by male alone.
6. Eggs mouth-brooded by male and young mouth-brooded by both parents.

The most complex spawning behaviour is to be observed in those species that mouth-brood both eggs and young. Permanent or temporary sexual dimorphism is common among cichlids, but may be absent. In some species that lay eggs on the substrate, the sexes may only be discernible during the spawning period, when the genital papilla is visible. In males it is conical and pointed, whereas in females it is cylindrical and of uniform diameter along its entire length. In the majority of species sexual dimorphism entails brighter coloration in the male but not in the female.

The Cichlidae can be difficult to keep with other aquarium fish, but the problem can sometimes be resolved by using a large tank, with a suitable environment, numerous hiding places and a careful selection of associated species.
☐
Aequidens curviceps, p. 38
Apistogramma agassizi, p. 40
Astronotus ocellatus, p. 40
Cichlasoma meeki, p. 45
Cichlasoma octofasciatum, p. 45
Etroplus maculatus, p. 47
Geophagus jurupari, p. 47
Haplochromis burtoni, p. 48
Hemichromis bimaculatus, p. 49
Lamprologus brichardi, p. 51
Papiliochromis ramirezi, p. 55
Pelatochromis kribensis, p. 55
Pseudotropheus zebra, p. 57
Pterophyllum scalare, p. 57
Sarotherodon mariae, p. 58
Symphosodon discus, p. 59
Tropheus moorei, p. 61

GOBIIDAE
Ord. Perciformes

Although the gobies are bottom-dwelling, often sedentary, fish, many species are interesting and excellent subjects for the aquarium. Their most noted feature is the pelvic fins which are fused together to form a suction disc whereby they can attach themselves to hard surfaces or settle on soft substrates. The fish are small, elongated and cylindrical, normally with two well-separated dorsal fins, the first supported by spines, the second by soft rays. The caudal fin, as a rule, is rounded. The mouth is terminal and the lips are prominent. The gill opening is small. The scales may be cycloid or ctenoid, sometimes absent. Normally they have no swim bladder, so they cannot float, but can remain undisturbed, even by currents, on the bottom.

The Gobiidae constitute the largest family of principally marine fish, with more than 1,500 species, which also include a few freshwater and brackish water forms. Most of them are tropical but there are a number of temperate species. Many live buried under the sand, in lairs or holes; others frequent rocky habitats, in cracks and crevices; and a few use other animals, such as sea urchins or coral, for shelter.

As a rule, the fish lay elongated eggs, furnished with peduncles, which affix themselves to rocks, coral or the walls of their burrow. Many species guard or tend the eggs until they hatch, and often both parents share this task.
☐
Brachygobius xanthozona, p. 43

ANABANTIDAE
Ord. Perciformes

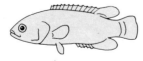

The labyrinth fish have a relatively compressed, high body, a large mouth and strong, conical teeth in both jaws. The dorsal and anal fins have a broad base and a number of robust spiny rays; the caudal fin is rounded and the ventral fins are in a thoracic position. The family, typical mainly of freshwater zones of Africa and Southeast Asia, is divided into three genera: *Anabas*, *Ctenopoma* and *Sandelia*. There are some 40 species in all, of which the best known for the aquarium is *Anabas testudineus*, the climbing perch. The principal feature of this family and its relatives (Belontiidae, Helostomatidae and Osphronemidae) is the presence of a supplementary breathing organ made up of many soft plates of densely vascularized tissue, situated inside the gill chamber. This organ, because of its shape, is called the labyrinth (hence the common name of these fish) and it enables the Anabantidae to breathe oxygen contained in atmospheric air and to live in poorly oxygenated water.

Given the presence of this accessory breathing organ, the aquarist is well advised to leave enough space between the water surface and the aquarium cover.
☐
Anabas testudineus, p. 38

BELONTIIDAE
Ord. Perciformes

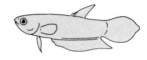

Similar to the Anabantidae, the Belontiidae have a protrusile upper jaw, short-based dorsal fins, long-based anal fins, and often pelvic fins with a long hair-like ray. The family is divided into ten genera and some 30 species, distributed from West Africa through India and Southeast Asia to Korea. They are generally easy to keep, but the aquarium must contain floating vegetation and, as a rule, have a dark bottom. Temperatures above 27°C (79°F) are acceptable for the tropical species, and below 20°C (68°F) for the temperate species. Although, as a rule, they feed on small live organisms, they adapt to dried food. In most species of Belontiidae, the male builds a bubble nest that may consist of plant material, beneath which he courts the female. The courtship may be elaborate, comprising displays, twists and turns, pursuits and bites, and culminating in an embrace whereby he wraps himself around her. If the eggs, which normally float, do not by chance end up in the nest, the male gathers them together and deposits them there, incubating them until they hatch a couple of days later. It is necessary to remove the female after egg laying and also the male as soon as the larvae become free-swimming. The larvae usually accept fish food in liquid or paste form but may need infusorians.
☐
Betta splendens, p. 42
Macropodus opercularis, p. 52
Trichogaster leeri, p. 60
Trichogaster trichopterus, p. 61

HELOSTOMATIDAE
Ord. Perciformes

Although they greatly resemble the members of the family Belontiidae, the Helostomatidae, represented by a single genus and one, perhaps two, species, have a different head structure and are characterized by a long-based dorsal fin which is almost symmetrical to the anal fin. The mouth is protractile, with thick, mobile lips that have earned these fish the common name of kissing gouramis. In fact, they are in the habit of pressing their lips against those of another individual. According to recent studies, this is merely a form of ritualized combat, a kind of challenge or threat for the defense of territory or the definition of hierarchy within a group.

The family is found in Southeast Asia, Thailand, Sumatra, Java, Malaysia and Borneo.
☐
Helostoma temmincki, p. 49

OSPHRONEMIDAE
Ord. Perciformes

The Osphronemidae contain only one genus and a single species, *Osphronemus goramy*, the giant gourami, probably the largest of Anabantoidea, measuring up to

60 cm (24 in) in length. The general shape of the body most resembles that of the Belontiidae, the difference being that the head of this fish is comparatively small and that in older individuals it has a distinctive "chin." Typical, too, are the large scales and the continuous lateral line. The ventral fins are reduced to thin, very long filaments which almost touch the caudal fin. This family was originally confined to the fresh waters of Borneo, Java and Sumatra.
□
Osphronemus goramy, p. 54

TETRAODONTIDAE
Ord. Tetraodontiformes

The fish of this order have fused incisor-shaped teeth. As a rule, they propel themselves by means of the pectoral fins or the dorsal and anal fins. The premaxillary and lower jaw are fused because the mouth is nonprotrusile. Large pharyngeal teeth are sometimes present. The scales are modified into spines or plates, and are occasionally absent. Many species produce sounds. The order is divided into eight families which include the puffer fish, the porcupine fish and the trigger fish, made up of 65 genera and about 320 species. The family Tetraodontidae (puffers) alone contains ten genera and more than 100 species. They possess a modified stomach that can swell up (normally by swallowing water); this characteristic permits the fish to increase their dimensions enormously so as to discourage predators. They have powerful jaws with fused teeth that form four beak-like incisors. The dorsal and anal fins have soft rays that are absent in the pelvic fins. The pectoral fins are linked to the dorsal and anal fins to provide greater propulsive thrust. By using the pectorals, the fish can swim backwards quite easily. The scales are sometimes modified into spines or are missing. The viscera, particularly the gonads, may contain a toxin.
□
Tetraodon fluviatilis, p. 60
Tetraodon palembangensis, p. 60

Families of tropical and temperate marine fish

About 60 per cent of all fish live in the sea. Some have a very localized distribution, but others are virtually cosmopolitan. They occupy an immense variety of habitats that range from coastal waters out to the open sea and down to the ocean depths, exhibiting a diversity of forms, colours and behaviour patterns that is a source of boundless interest and wonder.

TROPICAL WATERS

MURAENIDAE
Ord. Anguilliformes

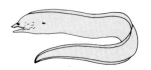

The moray eels have no pectoral and pelvic fins; the gill openings are small and round, and the lateral line is visible only on the head. There are no scales and as a rule the skin is smooth and leathery. All are marine species, most of them tropical, but several forms are to be found in temperate seas. There are 12 genera with about 100 known species.
□
Gymnomuraena zebra, p. 88
Gymnomuraena meleagris, p. 89
Muraena pardalis, p. 96
Rhinomuraena amboinensis, p. 105
Muraena helena, p. 117

CONGRIDAE
Ord. Anguilliformes

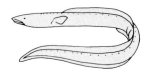

Some of the conger eels belonging to this family look, from a distance, like plants swaying in the wind. They live in lairs in the sand or other soft substrates, emerging partially to face the current and feed on the benthic plankton that is swept along. At night, or when approached, they withdraw completely into the sand for protection. Even when they couple, they emerge only far enough to intertwine their bodies and bring their bellies together for egg laying. They often form large colonies, some of which, situated in the Gulf of California, contain millions of individuals. There are four known genera and eight species of congers; the entire family comprises about 100 species.
□
Taenioconger digueti, p. 108

PLOTOSIDAE
Ord. Siluriformes

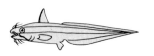

The Plotosidae are oblong, eel-like fish, with barbels around the mouth. The caudodorsal fin is very long and forms a single entity with the caudal and anal fins. Pointed and toothed spines, capable of causing serious wounds, are situated above the first dorsal fin. There are no scales. Some 30 known species are found in the Indian and Pacific Oceans, both in salt and brackish water, and in adjacent freshwater streams and rivers. The fish live together on the bottom, normally on soft substrates.
□
Plotosus anguillaris, p. 101

ANTENNARIIDAE
Ord. Lophiiformes

The first ray of the dorsal fin of all fish in the order Lophiiformes is modified so as to form a line with bait to attract prey to within a sufficient distance for them to be sucked in and swallowed. The members of this family have a spherical shape and loose skin, without scales but sometimes covered with small teeth. Some species, such as the sargassum fish, *Histrio histrio*, have patterns on their skin that resemble elements of their surroundings, so that they blend perfectly with the background. The gills are small and situated close to the pectoral fins. The Antennariidae move very slowly and can use their pectoral fins to "walk" on the bottom. There are about 60 known species which range in size

The division here of fish into freshwater/marine, tropical/temperate is not to be taken as a scientific or strict one; species of many families are found in both categories or may predominate in the counterpart. These outlines are used to maintain the distinctions used in earlier chapters and to describe the families of our selection of aquarium fish in the most accessible way.

from 3 cm (1¼ in) to 36 cm (14½ in). They are found in all tropical and subtropical seas.
□
Antennarius hispidus, p. 71
Antennarius sanguineus, p. 71
Histrio histrio, p. 90

OGCOCEPHALIDAE
Ord. Lophiiformes

These oddly shaped fish are poor swimmers but can "walk" on the bottom thanks to their specially adapted pectoral and pelvic fins which are similar to limbs. Their body is highly compressed and covered with tubercle-like scales; a retractile spine is all that exists of the first dorsal fin and is modified to form a rod and bait. There are about 55 described species divided into eight or nine genera. They live on soft bottoms in deep water and are seldom kept in aquaria.
□
Zalieutes elater, p. 121

ANOMALOPIDAE
Ord. Beryciformes

As a rule, the fish of this order are looked upon as an intermediate group between the lower soft-rayed fish and the higher fish with well-developed fin spines. As might be expected, therefore, characteristics of both groups are apparent. Usually the first dorsal fin is large and spiny, partly separated from the second (soft) dorsal fin by a groove. The relatively large pelvic fins are positioned behind the pectoral fins. The scales are ctenoid or cycloid, and the swim bladder is open or closed (sometimes absent). They live in all oceans and various habitats, from shallow tropical waters to deep seas. There are some 140 known species in 40 genera within the order, grouped into 15 families.

This small but distinctive family comprises four species, each of which possesses a large, kidney-shaped luminous organ (containing luminescent bacteria) underneath both eyes. These light organs are considered unique inasmuch as they can be extinguished either by raising a flap of skin to cover them (*Photoblepharon*), by rotating them inside (*Anomalops*) or by both methods (*Kryptophanaron*).
The Anomalopidae live in the darkness of deep water or in caves. They are highly sensitive to light and swim in surface waters only on dark nights. Apparently they use their special organ to avoid predators and to communicate with one another. Because they flee when approached by light, scuba divers were never able to see them until they discovered that the fish could be identified on dark nights just by their luminous organs. At this point, if the diver switches on his torch, momentarily freezing the fish, he can catch it with a net. Unfortunately, *Anomalops katoptron*, the species most often available to aquarists, is also the least hardy.
□
Anomalops katoptron, p. 71
Photoblepharon palpebratus, p. 101

MONOCENTRIDAE
Ord. Beryciformes

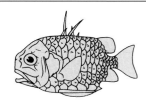

The fish of the family Monocentridae are much esteemed by marine aquarists for their appearance and their adaptation to captivity. They are stocky, oblong and slightly compressed; the mouth and eyes are large and they are covered by bony scales with a sharp central spine.
□
Monocentris japonicus, p. 96

HOLOCENTRIDAE
Ord. Beryciformes

These fish have a fairly compressed, high body, normally reddish in colour, with big eyes, strong and pointed fin spines and sharp scales. They are essentially nocturnal and spend the day inside caves and crevices in shallow waters off rocky coasts, generally at less than 30 m (100 ft).
Many of them are excellent aquarium subjects. Hardy and active, they soon get accustomed to feeding by day; they are not aggressive towards their own kind or other species, nor are they often attacked in their turn. The 70 known species are all tropical marine forms, living in the Atlantic, Pacific and Indian Oceans, the majority being found in the Indo-Pacific region.
□
Adyorix suborbitalis, p. 69
Holocentrus rufus, p. 91
Myrypristis murdjan, p. 97

CENTRISCIDAE
Ord. Syngnathiformes

This order includes the pipefish, sea horses and trumpet fish. All of them have a small mouth at the tip of the tubiform snout and a nonprotrusile upper jaw. The shrimpfish of the family Centriscidae are very strange, both in structure and behaviour. Their body, oblong and highly compressed, is covered in scales which form a thin central margin like a razor blade. They have no teeth nor a lateral line. The first dorsal fin is positioned terminally. The fish swim vertically with the head downward, hiding and camouflaging themselves perfectly among spiny sea urchins or algae. They are both interesting and rare. Only four species are known, all distributed in the tropical Indo-Pacific region and from East Africa to Hawaii.
The fish of the family Macrorhamphosidae are fairly similar in looks and behaviour, though with a more conventional body structure.
□
Aeoliscus strigatus, p. 69

SYNGNATHIDAE
Ord. Syngnathiformes

The pipefish have an oblong S-shaped body, plated with bony rings, and a tubiform snout. The gills and the anal fin are very small. There is a single dorsal fin and no pelvic fins. The eggs are incubated by the male in a ventral sac.
There are about 150 species, some two dozen of which are sea horses. The majority of species are found in tropical and temperate zones, shallow sea or brackish waters, although some inhabit fresh water. The sea horses are tropical marine forms, rarely living in temperate regions. The largest is the giant sea horse from the Pacific, measuring 30 cm (12 in).
□
Dunkerocampus dactyliophorus, p. 85
Hippocampus kuda, p. 90
Hippocampus hudsonius, p. 116
Syngnathus leptorhynchus, p. 120

SCORPAENIDAE
Ord. Scorpaeniformes

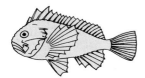

The fish of this order are notable for a bone known as the suborbital support, which extends from below the eye across the cheek. The body shape varies from cylindrical to compressed; the head and body are often covered with spines; the pectoral fin is rounded and the caudal fin round or square. The order contains 21 families, 260 genera and about 1,000 species. The members of the family Scorpaenidae may be active or sedentary, the latter often masters of disguise. The head is large and frequently covered with spines; the mouth is normally

broad; the pectoral fins are almost always large and rounded; and the tail is round or square. Fertilization is internal; their development may be external (oviparous) or internal (ovoviviparous). Approximately 325 species of the family are known.
☐
Dendrochirus zebra, p. 84
Pterois volitans, p. 104
Scorpaenodes xyris, p. 105
Scorpaena guttata, p. 119
Scorpaena scrofa, p. 119
Sebastes rosaceus, p. 120
Sebastes rubrivinctus, p. 120

SYNANCEIIDAE
Ord. Scorpaeniformes

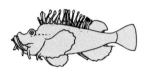

This family comprises the world's most venomous fish. Because of the danger they pose, they are not recommended for home aquaria (although there is an anti-venom against their poison). Importation and possession of these species are subject to regulations which vary from country to country and should be verified.
The body of these fish is more or less oblong and of normal appearance, but the head, which is covered with crests, spines and hollows, looks grotesquely deformed. The pectoral fins are very large and the single dorsal fin bears pointed anterior spines; the venom glands are situated at the base of each spine. Both scales and swim bladder are absent. The fish have camouflage coloration and normally rest partially submerged in the mud or among rocks and coral. Because they are usually found in very shallow water, it is easy to tread on them when wading or collide with them when swimming. Their wound is very painful, often causing difficulties in breathing and sometimes death.
There are about 30 species, all marine and living in brackish water; they are found from South Africa to the Society Islands, including the Red Sea and Australia. They are not present in Hawaii and the eastern Pacific.
☐
Synanceia verrucosa, p. 107

SERRANIDAE
Ord. Perciformes

The sea basses and groupers are sturdy fish with a large mouth, strong teeth and a fairly compressed, high body. The spiny dorsal fin is attached to a part with soft rays, with or without a groove. The caudal fin may be rounded, truncate, moon-shaped or slightly bifurcate. The mouth is subterminal and the lower jaw projects beyond the upper one. The scales are normally ctenoid, the lateral line is complete and the swim bladder is closed.
The Serranidae are all carnivorous and most of them live in the depths. Many are capable of quick colour changes. The fish are hermaphrodites, normally in sequential stages, first as male and then as female. As adults they may grow to a considerable size and thus are of little interest to aquarists. There are some 370 species, varying from about 10 cm (4 in) to over 3 m (10 ft) with weights of more than 400 kg (680 lb). The majority of species inhabit tropical seas, but some are found in temperate warm seas, and a very few live in temperate cold waters. There are also several freshwater species.
☐
Cephalopholis argus, p. 75
Cephalopholis miniatus, p. 76
Cromileptes altivelis, p. 83
Epinephelus dermatolepis, p. 85
Hypoplectrus unicolor, p. 92
Liopropoma rubre, p. 94
Serranus fasciatus, p. 106
Serranus tabacarius, p. 106
Epinephelus guaza, p. 116
Serranus cabrilla, p. 120

GRAMMISTIDAE
Ord. Perciformes

These close relatives of the Serranidae produce a protective mucus. There are some 17 species, found in the Atlantic and Indo-Pacific region.
☐
Grammistes sexlineatus, p. 88

PSEUDOCHROMIDAE
Ord. Perciformes

These small and active fish are closely related to the Serranidae but differ from them principally by their interrupted lateral line. The family consists of many species suitable for keeping in the aquarium, variously coloured.
☐
Pseudochromis flavivertex, p. 103
Pseudochromis porphyreus, p. 104

TERAPONIDAE
Ord. Perciformes

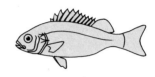

The Teraponidae are similar to the Serranidae, but differ in certain anatomical details. They are found only in the Indo-Pacific region where they live in salt, brackish and fresh water. Many specimens are striped. At least 39 species are known.
☐
Terapon jarbua, p. 108

KUHLIDAE
Ord. Perciformes

The Kuhlidae have a fairly high, compressed, silvery body, large eyes, an oblique mouth, deeply toothed dorsal fins and a bifurcate tail. They live in various habitats: coral reefs, mangroves, sea and fresh water. Some species shelter under jutting rocks. The fish are used as food and for bait.
Little is known about their reproduction but it is thought that eggs are laid throughout the year and that the young inhabit shallow water. There are about 17 species from East Africa and the tropical zones of the Indian and Pacific Oceans.
☐
Kuhlia taeniura, p. 93

APOGONIDAE
Ord. Perciformes

The Apogonidae are small nocturnal fish with colours that range from reddish to bronze. The head, eyes and mouth are large, the two dorsal fins are completely separated and of similar size, and the large, prominent scales are ctenoid or sometimes cycloid. The largest species measure up to 20 cm (8 in), but the majority are 10 cm (4 in) or even less. There are about 192 species divided into 26 genera, inhabiting tropical zones. Almost all are marine, living in shallow water in and around coral reefs; some, however, are found in rock pools and a few in deep water. By day they hide in dark places such as caves and cracks, gathering in groups; at night they come out and disperse to feed on plankton. Some species seek refuge in other organisms such as sponges and gastropods. As a rule, these fish are mouth-brooders. According to species, eggs and young are cared for by the male, by the female or by both parents jointly. Apparently they are short-lived and some are literally annual. As a rule, they are tranquil aquarium subjects, feeding readily. They need hiding places and are most active at night. In the wild they can be caught by scuba divers with hand nets.
☐
Apogon maculatus, p. 71
Apogon retrosella, p. 72
Cheilodipterus macrodon, p. 80
Paramia quinquelineata, p. 100
Sphaeramia nematopterus, p. 106

HAEMULIDAE
Ord. Perciformes

The family Haemulidae comprises several species which some authors classify in the families Pomadasyidae, Pristipomatidae, Gaterinidae or Plechtorhynchidae. They differ from the Lutjanidae in their dentition, mainly because they lack large canines. In many species the inside of the mouth is brightly coloured and is displayed during courtship or defense of territory. They grind their pharyngeal teeth, producing sounds that are amplified by the swim bladder. Some species have conspicuous, abnormally large lips. As a rule, the young differ much from the adults, with brighter colours and streaks. In the daytime they live in groups close to reefs and at dusk move down to the seabed, hunting shrimp, bivalves, worms and other invertebrates in the soft substrate and the vegetation. There are about 175 species, divided into 17 genera. Only those species whose young are brilliantly coloured are of interest for the aquarium.
☐
Anisotremus taeniatus, p. 70
Haemulon sexfasciatum, p. 89
Plectorhynchus chaetodonoides, p. 101

LUTJANIDAE
Ord. Perciformes

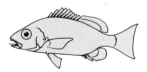

The Lutjanidae are medium-sized fish, seldom exceeding 1 m (3 ft) in length, and similar to the Serranidae. The body is tall or oblong, more or less compressed. Some species, such as those of the genus *Caesio* (sometimes placed in a separate family, Caseionidae) are tapered in shape. The dorsal fin is continuous and has strong spines; the caudal fin is bifurcate or slightly forked. The body is covered with average or large ctenoid scales. The lateral line is complete. The operculum is normally closed and sometimes has a sharp edge.
The Lutjanidae usually differ from the Serranidae in that they have a longer snout, an inclined forehead and a thin upper lip, this feature being due to the fact that the bone of the upper jaw (maxillary) is situated under the zygomatic or cheekbone. As a rule, there are large, strong canine teeth.
Most of the Lutjanidae live on the bottom, hunting fish or crustaceans at dusk or during the night, except for the *Caesio* species which feed on plankton. As far as is known, the fish lay pelagic eggs. The young of many species are often found close to shore and differ from the adults in colour and pattern. Some species raise their young in mangrove swamps. The family is distributed in warm seas. There are some 230 species in 23 genera. Many species are brightly coloured, but only a few are of interest to aquarists because the adults grow so big.
☐
Lutjanus sebae, p. 95

PRIACANTHIDAE
Ord. Perciformes

The representatives of this family have very large eyes; they are generally reddish but can change colour quickly, passing from a uniform tint to one mingled with streaks, stripes and spots. The Priacanthidae are nocturnal by habit and are found in temperate waters all over the world, normally near the bottom.
☐
Pristigenys serrula, p. 118

SCIAENIDAE
Ord. Perciformes

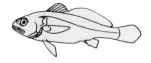

The Sciaenidae are fish of average length, slightly compressed and generally silvery as adults. Some species have a tapered shape. They are easily distinguishable from other fish by their lateral line, extending the whole length of the body above the tail. The majority of the Sciaenidae have their mouth in the lower part of the body, an adaptation which permits them to feed in the depths; some, however, have a terminal mouth with protractile jaws. Many possess a single barbel or a group of small barbels beneath the chin; and like the Lutjanidae, some have a thin upper lip. The dorsal fin is deeply toothed and sometimes completely divided; the caudal fin is rounded or truncate. The scales may be cycloid or ctenoid. Some Sciaenidae are characterized by a modified swim bladder, with many ramifications, to amplify sounds. Resonance is achieved by means of muscles situated on the walls, and the sound given out resembles a croak or a rumble. In species that lack a swim bladder, the sound is emitted by grinding the pharyngeal teeth. These noises may be used either for navigation or communication. There are about 210 known species divided into 50 genera, found mainly in warm waters. They are not present in Hawaii. Many species live principally in estuaries and some in fresh water. The young are suitable for the aquarium.
☐
Equetus lanceolatus, p. 85
Pareques viola, p. 100

GRAMMIDAE
Ord. Perciformes

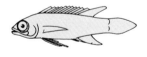

These small, highly colourful fish are near relatives of the Serranidae, differing by the lateral line, which is interrupted or absent. There are some 13 species from the tropical western Atlantic and Indo-Pacific region.
☐
Gramma loreto, p. 88

EPHIPPIDIDAE
Ord. Perciformes

The Ephippididae have a very compressed body and look much like the Chaetodontidae, but differ from these in that the dorsal fin is generally divided and is only partially spiny. The adults, as a rule, are silver with black stripes; the young vary from copper to reddish. It seems that some young are mimics, taking on the appearance of a dead leaf or other plant so as not to be spotted by predators, and that others use their black stripes to hide among the spines of sea urchins. The adults live in groups and can be found in a wide range of habitats, from deep reefs to shallow, flat seabeds and mangrove zones. During the night they may be seen swimming close to harbours in search of invertebrates attached to jetties and piers. There are some 17 species in five genera, widely distributed in tropical waters, but several in temperate zones. In the aquarium they are tranquil, often quite hardy and easy to keep. They do not appear to have been bred in captivity.
☐
Chaetodipterus zonatus, p. 76
Platax orbicularis, p. 101

MULLIDAE
Ord. Perciformes

The mullets are easily recognized by the pair of long barbels beneath their chin. They have an oblong, slightly compressed body covered by large ctenoid scales. The lateral line is complete. The head is blunted, the terminal mouth is positioned ventrally and the jaws are protractile with small conical teeth. The two dorsal fins are set well apart, the first being spiny and the second soft-rayed; the caudal fin is bifurcate. The Mullidae live on the bottom, preferably where the substrate is soft. They use their sensory organs, which contain taste buds, for locating

the small invertebrates on which they feed. Most mullets are vividly coloured and may undergo rapid colour changes under the influence of light and emotional stress. It appears they resort to this change in order to help cleaner fish to find external parasites.
Little is known of their reproductive behaviour save that the eggs and young are pelagic. If a light is shone at night, the young are attracted to it and can be caught with nets to be raised as aquarium subjects. There are some 55 species in six genera, very common in tropical seas, especially shallow water. Many mullets will adapt to aquarium life but they may cause problems by their habit of continually stirring up the sand. As a rule, they are delicate fish.
□
Mulloidichthys dentalus, p. 96

CHAETODONTIDAE
Ord. Perciformes

The colourful and active Chaetodontidae, the butterfly fish, are among the most popular of aquarium subjects. The body is high, oval and much compressed, and the small mouth contains bristle-like teeth. The dorsal fins are continuous, sometimes with a groove, and supported by both spiny and soft rays. The body is covered with ctenoid scales which are smaller on the head and extend to the base of the soft dorsal and anal fins. Young and adults are generally similar in colour. They are distinguished from the Pomacanthidae by the absence of an opercular spine and the presence of a distinctive "armoured" larval phase.
More than 100 species, divided into ten genera, have been described. They live in all tropical seas, particularly above the 20-m (65-ft) mark. Some species are to be found in deeper water, down to at least 200 m (650 ft), and yet others in temperate waters. Although the majority of butterfly fish inhabit coral reef zones, some prefer stony slopes and others rock walls. They live alone, in small groups or in large shoals.

The Chaetodontidae are often spectacular but the reason for their brilliant coloration is not wholly understood; it is thought that it may facilitate recognition, be used for courtship, in defense of territory or as a warning to predators.
Apart from being omnivorous, the fish have different feeding habits and four principal types have been described: hard coral, soft coral, benthic invertebrates and zooplankton. In addition, some may behave as cleaners, removing parasites or dead tissue from other fish. Certain species, when spawning, form pairs, which suggests temporary monogamy. Little, however, is known about their courtship or egg-laying habits. Groups of eggs, often in groups of 3,000 – 4,000, float to the surface where they develop rapidly, hatching within a day or so. The larvae feed on plankton.
Many species of Chaetodontidae can be raised successfully in the aquarium. Some, nevertheless, are difficult to keep because of their specialized feeding habits (e.g. coral polyps) and their aggressiveness or timidity. The species described in this book have been chosen for their beauty, availability and suitability for home aquaria; several of them are not commonly raised but are of interest to those aquarists looking for something unusual. The omnivorous species, or those that prey on crustaceans, are the easiest to keep; those that feed on sponges or algae are more difficult, and those whose diet is confined to coral polyps do not generally live long. In nature the fish eat continuously, so in the aquarium they should be fed at least twice a day, preferably three or four times. The majority adapt to temperatures of 20° – 24°C (68° – 75°F); those from temperate zones or deep water prefer 15° – 20°C (59° – 68°F); and those from warmer regions need temperatures of above 24°C (75°F). When higher than 28° – 30°C (82° – 86°F), however, oxygen solubility is low, and care must be taken to ensure that there the oxygen supply is sufficient.
□
Chaetodon aculeatus, p. 76
Chaetodon argentatus, p. 77
Chaetodon auriga, p. 77
Chaetodon collare, p. 77
Chaetodon ephippium, p. 77
Chaetodon falcifer, p. 78
Chaetodon humeralis, p. 78
Chaetodon kleini, p. 78
Chaetodon lunula, p. 78
Chaetodon ocellatus, p. 79
Chaetodon ornatissimus, p. 79
Chaetodon paucifasciatus, p. 79
Chaetodon rafflesi, p. 79
Chaetodon tinkeri, p. 80
Chaetodon ulietensis, p. 80

Chelmon rostratus, p. 80
Forcipiger flavissimus, p. 86
Hemitaurichthys polylepis, p. 89
Heniochus acuminatus, p. 90
Johnrandallia nigrirostris, p. 92

POMACANTHIDAE
Ord. Perciformes

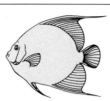

The Pomacanthidae are almost as popular with aquarists as the Chaetodontidae, even though adults of the larger species have to be accommodated in aquaria of 2,000 liters (440 gal) or more. They are distinguished from the other family by the presence of a strong preopercular spine. The young often have a different coloration to the adults, especially in the *Pomacanthus* and *Holacanthus* genera. There are about 75 species inhabiting tropical seas, particularly along the western coasts of the Pacific. They are generally found in shallow water, between the surface and a depth of 15 m (50 ft), close to rocks and coral reefs; but some are found in deeper water, down to at least 60 m (200 ft). There are three principal forms of diet: sponges, algae and plankton. However, this is normally mingled with benthic and encrustant organisms such as bryozoans, tunicates and polychaetes. Little is known of their reproductive behaviour. In some species the male has a harem composed of several females. The eggs are normally laid on the bottom, but they float to the surface and hatch within 18 – 30 hours.
□
Apolemichthys arcuatus, p. 72
Centropyge aryi, p. 74
Centropyge bispinosus, p. 74
Centropyge flavissimus, p. 75
Centropyge loriculus, p. 75
Centropyge potteri, p. 75
Chaetodontoplus mesoleucos, p. 80
Euxiphipops navarchus, p. 86
Holacanthus bermudensis, p. 90
Holacanthus ciliaris, p. 91
Holacanthus clarionensis, p. 91
Holacanthus passer, p. 91
Pomacanthus arcuatus, p. 102
Pomacanthus imperator, p. 102
Pomacanthus paru, p. 103
Pomacanthus zonipectus, p. 103
Pomacanthus diacanthus, p. 104

POMACENTRIDAE
Ord. Perciformes

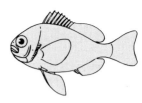

The Pomacentridae are laterally compressed marine fish. They possess both spiny and soft fin rays; they have ctenoid scales and a single nostril on either side of the head. The tail is normally bifurcate; it may, however, be rounded in certain *Amphiprion* species and truncate in *Dascyllus* species. Some species live in temperate waters and few grow longer than 30 cm (12 in). The majority of the more than 200 known species inhabit the zones around coral reefs in the tropical western Pacific.
The females lay eggs in nests prepared and watched over by the males. It is usually a simple cleaned area beside a rock or stone, into which the male drives the female and fertilizes the eggs she has laid. A single male may couple with several females at different times. Many species are highly quarrelsome in captivity and therefore unsuitable for aquarists. Some, however, like *Chromis*, *Dascyllus* and *Amphiprion*, are far less aggressive and have turned out to be excellent aquarium subjects, either alone or together with others. Food does not constitute a major problem because most species adapt quickly to a varied diet ranging from meat and fresh seafood to normal aquarium food.
The *Amphiprion* species are the clown fish, so called because of their brilliant colours and spectacular patterns. They are particularly interesting by reason of their symbiotic relationship with sea anemones, which normally eat fish. Of the 26 known clown fish species, 25 belong to the genus *Amphiprion* and one to the genus *Premnas*. All of them live in the Indian Ocean and the western Pacific. The *Amphiprion* species, small and medium sized, feed on zooplankton (and often algae as well) during the day, close to the anemones that act as their hosts; at night they withdraw into the anemone's tentacles. They are neither stung nor killed by the anemone's poisonous nematocysts because, so it seems, the mucus they secrete acts as an inhibitor. One anemone normally plays host to a pair of adult fish but may also provide shelter for one or two young. The host sea anemones comprise species of the genera *Cryptodendrum*, *Parasicyonus* and *Physobrachia*, but

mainly *Radianthus* and *Stoichactis*.
Eggs are laid all year round on a
rocky surface near the anemone.
Each female lays 300–700 eggs which
are watched over by both parents; it
is the male, however, who takes the
greater responsibility. The eggs hatch
within six to ten days and the larvae,
during their early life, are pelagic.
The genus *Dascyllus* comprises seven
species of small and medium-sized
Pomacentridae, many of which are
familiar aquarium fish. They are
tranquil species which learn to
associate with others but need plenty
of space and hiding places among the
rocks and coral. They come from all
the Indo-Pacific regions and include
D. albisella from Hawaii. The
Dascyllus species feed on zooplankton
and live in shoals on the barrier
reefs. Their diet also includes algae.
If threatened, they retire into the
reef. Some *Dascyllus* species,
especially when young, have
symbiotic associations with sea
anemones, similar to those from
which the *Amphiprion* species
benefit, but they depend on them less
for their survival. The eggs are laid
on a rocky surface or on dead
branches of coral and are guarded by
the male until they hatch, usually
within three days. The larvae spend a
short period as surface plankton prior
to settling on the reef. The young can
be raised in small 60-liters (13-gal)
aquaria although larger tanks are
preferable. Optimal water
temperatures are probably 24°–26°C
(75°–79°F). They can be fed with a
normal dried diet mixed with
vegetable matter.
☐
Abudefduf troschelli, p. 68
Amphiprion ephippium, p. 69
Amphiprion ocellaris, p. 70
Amphiprion perideraion, p. 70
Chromis atrilobata, p. 81
Chromis caerulea, p. 81
Chromis cyanea, p. 81
Chromis limbaughi, p. 82
Chromis scottii, p. 82
Chrysiptera cyanea, p. 82
Dascyllus aruanus, p. 83
Dascyllus melanurus, p. 84
Dascyllus trimaculatus, p. 84
Hypsypops rubicundus, p. 92
Stegastes flavilatus, p. 106
Stegastes rectifraenum, p. 107

CIRRHITIDAE
Ord. Perciformes

This small family of fish contains about 35 species in nine genera, the majority of which live in the Indian and Pacific Oceans. They are small, somewhat sedentary, predators which settle on rock or coral, waiting until their small victims approach and swallowing them whole. The family is notable for cirri on the margins of the nostrils and on the spiny part of the dorsal fin. The pectoral fins are normally adapted to allow the fish to rest on them: in fact, the lower rays are enlarged, unramified, often separated from one another and digitiform. Many species rest on branches of coral or *Gorgonia* and often mimic their surroundings. They are very agile in escaping traps and are sometimes caught by ripping off entire branches of corals and sea fans, a destructive practice which is at all costs to be avoided. The Cirrhitidae are for the most part excellent aquarium subjects, but it is best not to keep them with other small fish because they may eat them.
☐
Cirrhitichthys oxycephalus, p. 82
Oxycirrhites typus, p. 99

POLYNEMIDAE
Ord. Perciformes

The Polynemidae are characterized by a subterminal mouth and long, detached pectoral fin rays. It is thought that these separate rays contain tactile receptors and chemicals which are used to locate prey on soft bottoms. The family comprises seven genera and some 35 species, some of which grow to a length of almost 2 m (6 ft). It is widely distributed in tropical and subtropical salt and brackish waters. Only the young of the smaller species make good aquarium fish.
☐
Polydactylus oligodon, p. 102

LABRIDAE
Ord. Perciformes

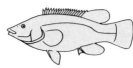

The Labridae, or wrasses, are among the most conspicuous and abundant fish of the coral reef. More than 500 species are known the world over, all of them marine and the majority from tropical zones; but some, found also in temperate seas, show an enormous diversity, ranging from a slim, cigar-shaped body to a sturdy, high body, and varying in length from 5 cm (2 in) to 1 m (3 ft). They are notable for their strange, jerky swimming movements, explained by the fact that they are propelled mainly by the pectoral fins rather than the caudal fin. When moving fast, however, they obtain additional thrust by using the tail as well. The lips are thick and protrusile, with almost separated teeth. There are rounded canines, which may be tusklike, in the front part of the mouth. The Labridae are diurnal. At night some hide for protection in the sand, others in crevices. The majority feed on benthic organisms but the family also contains species that consume plankton and detritus; some, too, are predators and others cleaners. In many species the young males and females are differently coloured. Sexual inversion (usually female to male) is common, and the inverted subject takes on a very special colour. Among many species, inverted individuals play a dominant role in reproduction, inasmuch as they often initiate egg laying in groups, although sometimes they will lay individually with selected partners. Many variations of sexual physiology and behaviour have been observed. Although many of the Labridae adapt to captivity, a careful choice must be made because some are very quarrelsome. They must be kept in an aquarium with a thick layer of sand, well aerated with an undergravel, preferably reverse-flow, filter. Inside the tank it is necessary to arrange rocks, lairs and crevices for those species that need to hide at night and shelter by day.
☐
Anampses meleagrides, p. 70
Coris gaimardi, p. 83
Gomphosus varius, p. 88
Hemipteronotus pavoninus, p. 89
Labroides dimidiatus, p. 93
Lienardella fasciata, p. 94
Thalassoma lucasanum, p. 108
Coris julis, p. 115
Labrus viridis, p. 117
Thalassoma pavo, p. 121

OPISTHOGNATHIDAE
Ord. Perciformes

The Opisthognathidae have a big mouth and large eyes and live in burrows that they build themselves. The body is more or less cylindrical, tapering slightly towards the tail. The large mouth with its conspicuous upper jaw is in a terminal position; both jaws possess average-sized canines. The dorsal and anal fins are long and continuous, without grooves; the pelvic fins are placed anterior to the pectoral fins; the tail is usually rounded. The scales are cycloid and absent on the head. The lateral line is high on the body and extends only about halfway to the tail.
These fish are generally found at depths of more than 4 m (13 ft), in sandy and muddy sediment mixed with stones, shells, etc. They build their burrow, which opens into a chamber, meticulously, with shells and stones that they grasp in their jaws; at night they use a stone or a shell to seal the entrance. Foreign bodies that fall into the lair are carried to the entrance and pushed outside. The fish feed on crustaceans and sometimes on other fish which stray too close. During the courtship period the male loiters above his burrow and performs a nuptial dance to attract the female. After laying the eggs, the female returns to her burrow and the male mouth-broods the eggs until they hatch. There are about 70 species, principally tropical. It appears they are not found in the western Pacific.
☐
Opisthognathus aurifrons, p. 98
Opisthognathus punctatus, p. 98
Opisthognathus rhomaleus, p. 98

SCARIDAE
Ord. Perciformes

Because the Scaridae, or parrotfish, are so conspicuous on tropical reefs, they are particularly interesting to aquarists. However, being so large and needing such a specialized diet, they are difficult to keep, even in large public aquaria. Anyone wishing

FAMILIES • 147

to try is advised to choose young fish which adapt more quickly to captivity. They are sturdy and generally large, ranging from 45 cm (18 in) to more than 2 m (6 ft). The body may be slightly compressed, high or relatively high. Many species display a marked protuberance on the forehead. The scales are cycloid and very large. The dorsal and anal fins are long; the dorsal fin is continuous. The mouth is terminal, and the maxillary teeth are fused to form a stout beak similar to that of a parrot; the pharyngeal teeth are strong and suitable for grinding coral, coral algae, etc. Some 70 species, all marine, are known.
☐
Scarus ghobban, p. 105

CHAENOPSIDAE
Ord. Perciformes

The Chaenopsidae are fish with a long, compressed body, without scales or a lateral line. The long dorsal fin is formed of spines and rays, separated or joined at the caudal fin; the anal fin, too, is long. Sometimes there are cirri above the eye.
☐
Chaenopsis alepidota, p. 76
Emblemaria hypacanthus, p. 85

CLINIDAE
Ord. Perciformes

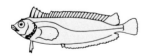

The Clinidae are similar and closely related to the Blenniidae, differing from them in that the scales are generally shiny and the maxillary teeth are linked in groups and conical, rather than in a line like a comb. They are fairly long, compressed fish with a typically pointed snout; there are cirri on the head and, characteristically, on the eyes and neck. The dorsal fins are normally continuous, with more spines than rays. The first spines may be partially or completely separated from the rest of the fin by a groove. The pelvic (ventral) fins are positioned in front of the pectoral fins. The majority of Clinidae have camouflage coloration. They are shy and solitary, living along rocky coasts from the surface down to some 30 m (100 ft) in depth, although some may be found at less than 10 m (33 ft) and others at more than 40 m (135 ft). Those species whose feeding habits are known are carnivores. The Clinidae that usually live among vegetation blend perfectly with the background and some adopt special techniques for camouflage. *Heterostichus rostratus*, for example, rests against a stem of a fugaceous plant and sways its body back and forth, mimicking a leaf, until an unwary fish approaches, at which point it makes a sudden leap and captures it. Some Clinidae are polychrome and may take on various colours that vary from red to brown or even to the same green as the vegetation of where it is living. Sexual dichromatism is also common, in which case the males and females have dark markings in different patterns.
The Clinidae are very widely distributed in tropical and temperate seas. There are some 75 species in 20 genera. As a rule, they are of little interest to aquarists because of their timid nature, which in some species becomes territorial and aggressive.
☐
Gibbonsia elegans, p. 86
Neoclinus blanchardi, p. 97

BLENNIIDAE
Ord. Perciformes

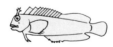

The blennies are small fish without scales and a stocky head. Most live in shallow water among rocks and coral, though a few are found in areas with strong wave action. Because many species are aggressive, aquarists tend to avoid the entire family. Generally the body is compressed and oblong, with soft skin, although in some species there are traces of scales. As a rule, there are cirri on the head. The eyes are positioned in the upper part of the head. The mouth is normally subterminal, with a series of sharp teeth in rows on the jaws, as in a comb. The dorsal fin is continuous but may be deeply grooved. A fundamental feature of the family is the position of the pelvic (ventral) fins, in front of the pectoral fins. The fish may be herbivorous or carnivorous. Most of them are cryptic and territorial. As a rule, the eggs are laid in shells or other forms of shelter and guarded by the male until they hatch. In species characterized by sexual dimorphism the males have brighter markings and sometimes special fins. Mimicry is also common; for example, some imitate cleaner fish, but instead of removing parasites from the hosts they attack them, ripping off scales or pieces of skin and flesh.
In tropical and temperate waters there are more than 300 species divided into 53 genera. In the western Atlantic only a few species are known. The Blenniidae are sturdy in captivity. Some are quickly domesticated, others are very aggressive in an aquarium and if kept with fish of different families, their territorial needs must be provided for.
☐
Aspidontus taeniatus, p. 73
Exallias brevis, p. 86
Hypsoblennius gentilis, p.92
Meiacanthus atrodorsalis, p. 95
Ophioblennius steindachneri, p. 98
Blennius ocellaris, p. 114
Blennius tentacularis, p. 115

GOBIIDAE
Ord. Perciformes

The gobies constitute the biggest family of principally marine fish, with more than 1,500 species. Many are easily recognizable by their pelvic fins which are fused to form a sucker; some do not have this but it is thought that they are evolved from fish that did possess it. The Gobiidae are small, almost all of them measuring less than 10 cm (4 in). The body is long, slightly compressed and with a large caudal peduncle. The head is usually round, with eyes on the upper part; the large, oblique mouth has small maxillary teeth. There are two dorsal fins, and the caudal fin is normally rounded but sometimes pointed. The scales may be cycloid, ctenoid or absent. There is no lateral line nor, as a rule, a swim bladder. The Gobiidae are carnivorous fish that live in the depths, often in direct contact with the substrate. Generally they remain resting on their pelvic fin close to a burrow or other refuge. Some species swim above the seabed and others live in symbiosis with various animals. Many form pairs. The eggs, as a rule, are laid in a hidden place and are watched over by the male.
Gobies live principally in tropical seas but they are also found in other watery environments, including fresh water. Many species are suitable for aquaria and a few have reproduced in captivity.
☐
Coryphopterus personatus, p. 83
Gobiodon citrinus, p. 87
Gobiosoma digueti, p. 87
Gobiosoma oceanops, p. 87
Gobiosoma puncticulatus, p. 87
Lythrypnus dalli, p. 95
Periophthalmus sp., p. 100
Coryphopterus nicholsi, p. 115
Gobius niger, p. 116

ACANTHURIDAE
Ord. Perciformes

The Acanthuridae, or surgeonfish, can be small, medium or large fish. They have a high, much compressed body and possess a typically sharp, retractile spine on each side of the caudal peduncle. Some, such as the *Naso* species, have two or three separated spines. The zone surrounding the spine is often brightly coloured. *Zanclus cornutus*, now included in the family, lacks spines. The small mouth is terminal and contains tiny maxillary teeth. The dorsal fin is long and continuous; the caudal fin may be truncate, bifurcate or slightly perforated with long appendages. The skin is leathery and covered with small scales. The lateral line is complete. The majority of species are diurnal and herbivorous, feeding in shallow waters on algae that live on rocks or coral. Some are omnivorous and a few consume zooplankton. The caudal spines are used in territorial or domination disputes and apparently as defenses against predators. The eggs are laid on shallow reefs and may involve either a pair or a group; they may either be pelagic or demersal and adhesive.

The Acanthuridae live in tropical waters all over the world. There are about 75 known species divided into ten genera. Most of them are popular in aquaria.
☐

Acanthurus achilles, p. 68
Acanthurus leucosternon, p. 68
Acanthurus lineatus, p. 68
Acanthurus triostegus, p. 69
Naso lituratus, p. 97
Naso unicornis, p. 97
Paracanthurus hepatus, p. 100
Prionurus punctatus, p. 103
Zanclus canescens, p. 109
Zebrasoma flavescens, p. 109
Zebrasoma veliferum, p. 109

SIGANIDAE
Ord. Perciformes

The Siganidae, like the Acanthuridae, have protective spines, but in their case, these are the spines of the anal and dorsal fins and they may be poisonous, causing painful stings. The body is high and compressed, the mouth is small or medium, and the teeth are adapted for browsing on plants. The fish, especially the young, may be brightly coloured or they can have uniform or camouflage colouring.
The majority live in shoals in shallow zones of the coral reef and in lagoons. The adults of some species live close to rock walls, to a depth of 30 m (100 ft) or more. It is a small family that comprises only two genera and about 25 species distributed in the tropical waters of the Indo-Pacific region and the eastern Mediterranean. Only one species is of interest to aquarists.
☐
Lo vulpinus, p. 94

CALLIONYMIDAE
Ord. Perciformes

The Callionymidae are small fish living in the depths, with large filamentous fins. The head is stocky and flat, with large eyes situated at the top. The mouth is small and terminal, with weak teeth; the gill openings are small. There are two dorsal fins. The caudal fin may be truncate, rounded or lanceolate, with visible rays. The Callionymidae live in habitats ranging from intertidal zones and low reefs to deeper waters. Many species hide during the day. Sexual dimorphism is common; the male has bigger fins. Fertilization is internal and the eggs are pelagic. These fish are found in tropical seas all over the world. There are some 130 species in ten genera. Many varieties, though sometimes delicate, are sought by aquarists for their beauty.
☐
Synchiropus picturatus, p. 107

BOTHIDAE
Ord. Pleuronectiformes

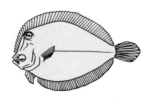

The members of the family Bothidae usually possess eyes on the left side of the head, which is turned upward. They are common on soft bottoms, at different depths, and live mainly in tropical and temperate seas. Most of them are carnivorous, feeding on small fish and crustaceans. There are more than 200 species divided into 37 genera.
☐
Bothus mancus, p. 74

BALISTIDAE
Ord. Tetraodontiformes

The Tetraodontiformes comprise fish that are furnished with unusual defensive systems. About three quarters of the known species belong to two families: Balistidae and Tetraodontidae. The Balistidae, or triggerfish, are equipped with a release mechanism that locks the thick spines of the first dorsal fin in an upright position, enabling them to shut themselves inside crevices to avoid capture. The body is high, compressed and wedge-shaped at the front, the leathery skin being covered with rough, square scales. The small mouth is terminal, with projecting incisors, and the eyes, too, are small. There are two dorsal fins; the second, situated far back, is symmetrically opposite the anal fin. The tail may be truncate, slightly forked or rounded. The Balistidae live in shallow coastal waters and are adapted to a diet of various hard-shelled animals, including sea urchins, crabs and coral. As a rule, they are solitary but some species form shoals, sometimes near the water surface. At night they enclose themselves in cracks for protection. Scuba divers can capture them by lowering the second dorsal spine so as to unlock the dorsal fin; at this point the fish can easily be pulled out of the hole. When swimming around the coral reef they use the dorsal and anal fins for locomotion. If they feel endangered, they utilize the caudal fin to gain additional propulsion.
Many Balistidae are edible but some species, particularly the large ones, can cause food poisoning if eaten. There are about 135 species, distributed in tropical zones of the Atlantic, Pacific and Indian Oceans. Some venture into temperate waters. Triggerfish are hardy, brightly coloured and very popular as aquarium subjects; but some are aggressive and therefore difficult to keep with other varieties.
☐
Balistapus undulatus, p. 73
Balistes vetula, p. 73
Balistoides conspicillum, p. 73
Melichthys niger, p. 95
Oxymonacanthus longirostris, p. 99
Rhinecanthus aculeatus, p. 105
Rhinecanthus rectangulus, p. 105
Sufflamen verres, p. 107
Xantichthys mento, p. 109
Balistes carolinensis, p. 114

OSTRACIIDAE
Ord. Tetraodontiformes

These are small fish, slow swimmers, with a box-shaped body encased in bony plates. The skin contains a toxic substance that is emitted when the fish is under stress; in a closed environment this may be fatal for the fish itself and for others. Predators generally avoid them, even though they live in open waters and are very conspicuous. To transport them, enclose each fish in a separate container, and when settling them into the aquarium, take precautions to avoid the poisoning of other inmates. In the species described here, the male is more colourful but less robust than the female. There are 30 known species in 13 genera, found in all tropical seas.
☐
Lactoria cornuta, p. 93
Lactoria fornasina, p. 93
Ostracion meleagris, p. 99
Ostracion tuberculatus, p. 99

TETRAODONTIDAE
Ord. Tetraodontiformes

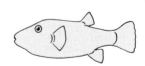

The puffer fish have a soft body and can swallow air or water to blow themselves up and ward off predators. They have strong jaws and teeth fused to form four incisors similar to beaks, two to each jaw. The dorsal and anal fins have only soft rays and provide the main propulsive power. Although they move slowly, they can maneuver skilfully and can easily swim backwards. The scales are modified into spines or are lacking. The viscera of certain species (e.g. of the genus *Fugu*) contain a deadly poison. The fish are found in tropical and temperate seas, in a wide variety of habitats. Some 118 species are known, divided into 16 genera, most of which have been kept successfully with other varieties.
☐
Arothron meleagris, p. 72
Arothron reticularis, p. 72
Canthigaster punctatissima, p. 74

DIODONTIDAE
Ord. Tetraodontiformes

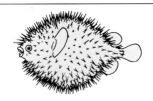

The Diodontidae have many elements in common with the Tetraodontidae but possess scales that are modified into spines and a single fused canine tooth in either jaw. They are widely distributed in tropical seas and sometimes in temperate waters. Fifteen species are divided into two genera. The smaller specimens are of interest, usually in aquaria containing a number of different fish.
☐
Diodon holacanthus, p. 84

CARCHARHINIDAE
Ord. Lamniformes

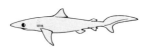

This is a large family of sharks, with species that vary in size and are very abundant in the tropics. It includes some species that pose a danger to humans. The Carcharhinidae are fusiform, with a pointed outline, two dorsal fins and a caudal fin with a broad upper lobe. The breathing apertures are small or absent. Most species are nomadic and viviparous. Those kept in captivity are normally active. Fertilization is internal. There are about 90 species; at least one, *Carcharhinus leucas*, is found in fresh water.
☐
Triakis semifasciata, p. 108

TEMPERATE WATERS

SCYLIORHINIDAE
Ord. Squaliformes

This family of sharks is found in both tropical and temperate seas. There are 20 genera and about 94 species.
☐
Scyliorhinus stellaris, p. 119

RAJIDAE
Ord. Rajiformes

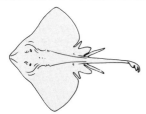

Sting rays, electric rays and skates are among the most common of the Rajiformes to be found in public aquaria. In all of them the body is clearly distinguished by a much larger anterior part (the disc), and a thin tail of variable length. The sides of the disc, rounded or pointed (in which case the outline is not round), are merely the large pectoral fins used for swimming; on the lower ventral side are the nostrils, the mouth and five pairs of gill slits; and on the dorsal side (or in a lateral position) are
the eyes, as well as two apertures called spiracles.
The rays have a rhomboid-shaped disc, the skin is covered by rough scales and the tail is fairly long and narrow. The females lay eggs that are enclosed in a horny oblong shell, with long tendrils extending from the corners. Some species are extremely large, their length exceeding 2.5 m (8 ft).
The Rajidae live mainly in temperate and cold seas and thus do not interest the tropical aquarist. There are 15 or so Mediterranean species, not always easy to identify. Recognition of the young is especially difficult, but they are the easiest to keep in small aquaria.
☐
Raja clavata, p. 118

TORPEDINIDAE
Ord. Rajiformes

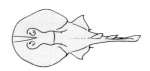

The torpedos or electric rays have a rounded disc, inside which are located two kidney-shaped electric organs (transparently visible in young individuals). The electric shocks are delivered from the ventral to the dorsal side and are controlled by the nervous system. They serve both as offensive and defensive weapons and were familiar to doctors in ancient times who used them for the first rudimentary form of electrotherapy. Torpedo rays are lazy, sedentary creatures, generally found on muddy or sandy bottoms. They are ovoviviparous and the embryos initially have a shark-like appearance.
☐
Torpedo marmorata, p. 121

DASYATIDAE
Ord. Rajiformes

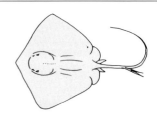

This family comprises various rays with a disc that may be either rounded or rhomboidal, and a tail furnished with a stout poisonous spine on the top. This particular appendage, with its inward-turned tip, is justly feared because it can cause painful stings; it is extremely sharp, with toothed margins that can aggravate the lacerations. Moreover, in two grooves on the lower side is a tissue secreting venom.
☐
Dasyatis pastinaca, p. 115

BATRACHOIDIDAE
Ord. Batrachoidiformes

The toadfish of the family Batrachoididae are sturdy and slow-moving, with a large head, wide mouth with strong teeth, and eyes situated on the back. As a rule, they have no scales but the body is well covered with mucus. There are one or two lateral lines. The first dorsal fin is short and supported by several strong spines; the second dorsal fin has a long base and is supported by rays. The pelvic fins are located in front of the pectoral fins. In some species the fins and opercular spines contain toxins that can cause painful wounds. The swim bladder is used to produce sounds, especially during courtship. They are principally coastal sea fish living on the bottom, in the tropical and temperate waters of the Atlantic, Pacific and Indian Oceans. Of the 64 or so known species, three are freshwater. The fish can live in poorly oxygenated water and survive for some time out of water.
☐
Opsanus tau, p. 117

COTTIDAE
Ord. Scorpaeniformes

The Cottidae are a large family of cold-water benthic fish similar to the Scorpaenidae. In general they are quite small, with a cylindrical body that narrows towards the tail, and a large head. As a rule, the eyes are situated high on the head; the mouth and lips are large. The dorsal fin is long, with the spiny part wholly or partially separated from the soft-rayed part. The anal fin is also long while the pectoral fin is large and fan-shaped; the tail is round or square. The scales may be totally absent or cover the body, in which case they are arranged in rows or groups and resemble bony plates or spines. Most species have cirri, some in large number. The lack of a swim bladder facilitates life on the bottom. Reproduction takes place in winter or spring and fertilization is internal;

only a few eggs are laid and they are guarded by the male.
The majority of species live in shallow water, but some are found deeper down and others (certainly anadromous) in fresh water. The Cottidae number about 70 genera and 300 species.
☐
Jordania zonope, p. 116
Nautichthys oculofasciatus, p. 117
Rhamphocottus richardsoni, p. 118

STICHAEIDAE
Ord. Perciformes

The Stichaeidae are small fish with a slender, compressed body. The terminal mouth is furnished with small conical teeth. The dorsal and anal fins are long, attached to or detached from the caudal fin. The dorsal fin is sharp to the touch. The pelvic fins are situated in front of the pectoral fins or are absent altogether. The scales, if present, are small and cycloid. With the exception of *Phytichthys chirus*, the fish tend to be brown, grey or black. Some species, for example *Chirolophis decoratus*, are provided with cirri, mainly on the head. These fish live on the bottom and in diverse habitats, from the rocky intertidal zone to muddy or sandy bottoms at a depth of more than 300 m (1,000 ft). Little is known of the reproductive behaviour of the deep-sea species. Those that live in shallow water lay bunches of eggs in nests placed under reefs. The eggs are guarded by one parent who fans them continuously to make sure the embryos receive enough oxygen as they grow and to keep the eggs free of sediment. There are about 60 species of Stichaeidae, found mainly in the cold waters of the North Pacific; a few varieties live in the North Atlantic. Some intertidal species adapt to warmer temperatures, around 20°C (68°F).
☐
Anoplarchus purpurescens, p. 114

PHOLIDIDAE
Ord. Perciformes

These are small fish with a long, slender body. The mouth is terminal, the teeth are small and conical, and the long dorsal and anal fins converge with the caudal fin. The pelvic fins, sometimes absent, are small and placed in front of the pectoral fins; the scales are tiny, cycloid and hardly visible. The coloration is predominantly green, yellow or red. Polychromatism may appear in the form of various colour phases within the same species. They are benthic fish that inhabit zones with rocky or soft bottoms where there is plenty of vegetation. They feed on small crustaceans and other invertebrates living among the plants. Spawning occurs in winter or early spring. The eggs, in bunches, are laid in nests placed below stones, and they are incubated by both parents until they hatch within 30 days at 10°C (50°F). There are some 13 species, mostly found in cold and temperate waters of the North Pacific; some live in the North Atlantic.
☐
Apodichthys flavidus, p. 114

SCYTALINIDAE
Ord. Perciformes

This family contains only one known species, a small worm-like fish which is found along the coasts of the northeastern Pacific. This strange individual lives among the gravel on stony beaches where the shore is partially sheltered. It is an unusual habitat and any other kind of fish would certainly be swept away or killed by the shifting gravel. It is an interesting, indeed unique, species for aquarists.
☐
Scytalina cerdale, p. 119

TRIPTERYGIIDAE
Ord. Perciformes

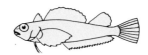

The Tripterygiidae are small fish with a dorsal fin divided into three parts: the first two are supported by spines, the third by soft rays. The body is slim and cylindrical, with the eyes situated near the top of the head, and the mouth subterminal. The tail is fan-shaped and slightly rounded. The pelvic fins are in a jugular position and attached in front of the pectoral fins. The anal fin is long. The scales are ctenoid. They are found in shallow waters and can sometimes be seen swimming about in rocky zones. They are normally camouflaged but can display conspicuous signals in order to communicate. They are all marine tropical forms from the Atlantic, Pacific and Indian Oceans. Some 95 species belong to 15 genera.
☐
Tripterygion nasus, p. 121

PLEURONECTIDAE
Ord. Pleuronectiformes

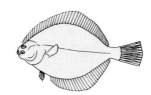

The fish of this order have a very flat body and lie on one side. As they grow, both eyes travel across to one side of the head, the fish resting on the eyeless side. The species of the family Pleuronectidae are mostly right-sided; however, as in the Bothidae, both left-sided and right-sided individuals may be found within the same species. They live mainly in cool waters and are distributed from the polar to temperate regions at varying depths. Some live in brackish water and a few in fresh water. There are about 100 species in 41 genera.
☐
Pleuronichthys coenosus, p.118

Disease guide

PRINCIPAL SYMPTOMS

Bold numbers following symptoms refer to possible diseases as listed in the following pages.

Changes in behaviour

- Agitated appearance: **1**
- Awkward movements combined with loss of colour: **12**
- Continuous rubbing against stones: **3, 4, 7, 8, 14, 17, 27, 29**
- Difficulty in breathing: **9, 27**
- Fins permanently folded: **8, 29**
- Gill covers move more rapidly: **4, 9**
- Problems of balance and loss of appetite: **21**
- Rapid breathing: **2, 7**
- Refusal to eat and fading of colours: **23**
- Slow, abnormal movements: **23, 24**
- Swimming jerkily: **16**
- Swimming with difficulty: **7, 15**

Changes in appearance

- Appearance of holes on the head: **21**
- Breakdown of rays: **27**
- Cloudy eyes: **4**
- Cottony, fluffy formations: **13**
- Deepening of colours: **21**
- Deformed backbone: **27, 28**
- Emaciated body: **9, 15, 16, 27, 28**
- Emaciated body, bulging eyes, skin changes: **25**
- Emaciated body, sunken eyes, pale gills: **3, 24**
- Flaking skin: **3**
- Frayed fins: **5**
- Gill covers remain open: **9**
- Inflamed epidermis: **27**
- Inflamed gills: **4**
- Inflamed ulcers: **7, 8, 14**
- Jagged fin edges: **27**
- Opaque fin edges: **27**
- Presence of ulcers: **5, 16, 21**
- Protruding eyeballs: **10**
- Raised scales: **10**
- Reddish, bleeding ulcers: **18**
- Shortening of fins: **27**
- Stunted growth: **28**
- Sunken eyes: **10**
- Swollen belly: **10, 15, 28**
- Whitish patches: **3, 4, 7, 14**
- Whitish slime on body, especially near mouth: **5**
- Whitish slime, white borders to scales: **5**
- Whitish ulcers: **11, 16**
- Widespread ulceration: **5, 10, 18**
- Yellowish brown or greyish coloration: **27**

External signs and parasites

- Black spots: **2**
- Growths of various shapes and sizes: **26**
- Long worm-like parasites or with egg sacs: **17, 19**
- Rounded parasites on body and fins: **1**
- Slow-growing red or whitish nodules: **20**
- Small greyish white nodules: **8, 15, 29**
- White spots: **25, 29**
- Whitish or pinkish cauliflower-like growths: **25**
- Worm-like parasites on body: **14**
- Worm-like parasites on gills: **9**

LIST OF MOST COMMON DISEASES

1. Anchor worm
2. Argulosis
3. Black-spot disease
4. Chilodonella
5. Cloudy eyes
6. Columnaris disease
7. Copepods
8. Costiasis
9. Cryptocaryoniasis or Marine white-spot disease
10. Dactylogyrosis
11. Dropsy
12. False neon fish disease
13. Fin rot
14. Fish pox
15. Gyrodactylosis
16. Ichthyosporidium disease
17. Internal flagellate disease
18. Internal worm diseases
19. Leeches
20. Lymphocystis
21. Neon fish disease
22. Saprolegnia-type fungal infections
23. Septicaemia
24. Sleeping sickness
25. Tuberculosis
26. Tumours
27. Velvet disease
28. Velvet disease of coral fish
29. White-spot disease

1. Anchor worm

This disease is caused by copepod crustaceans of the genera *Lernaea* (*Lernaea cyprinacea*, *L. elegans* and *L. phoxinacea*), known as anchor worms because the females (the actual carriers of the disease) have a head shaped like an anchor, with which they penetrate the musculature of the host fish, thus introducing the parasite into the body cavity and sometimes into the liver. These copepods measure up to 20 mm (¾ in) long and reproduce at temperatures above 14°C (57°F). The disease most frequently afflicts fish raised in open-air ponds, and particularly the Cyprinidae, although in the aquarium the Cichlidae tend to be singled out. The parasite is clearly visible on the skin of the affected fish, from which protrude elongated or oval filaments furnished with a pair of sacs that contain the eggs. The skin around the infected points may be reddened. The disease may appear in the aquarium after the purchase of plants or of fish already infected.
Treatment – Larger worms can be removed with tweezers, if great care is taken not to cause pain to the fish. Pay attention to wounds thus produced as they can take a long time to heal and become the source of secondary and much more dangerous infections.

2. Argulosis

This is a disease caused by crustaceans of the genus *Argulus*, better known as fish lice. They are fairly big parasites (8–10 mm/¼ in) and easily visible on the skin of fish, attaching themselves by hooked suckers. The mouth forms a long pointed proboscis, through which poison is ejected. They swim freely from one fish to another, feeding on their blood, and leaving red spots and pink patches surrounded by the surplus mucus secreted by the fish for defensive purposes. The wounds eventually turn necrotic and ulcerous, causing secondary infections.
Treatment – It is possible to remove the parasites, given their size, with a pair of tweezers after swabbing the fish with a strong salt solution, which also serves to disinfect the wound. The treated fish should then be placed in a quarantine tank. Alternatively, bathe with potassium permanganate (1 g/100 l) or formalin (1 g/500 l for 15 minutes). The larvae of fish lice survive for only 2–3 days without hosts. Proprietary treatments containing organophosphorus insecticides are available from aquarium shops and are very effective.

3. Black-spot disease

This disease is caused by the larvae of a worm such as *Posthodiplostomum cuticola* which burrow into the skin and fins of the fish, depositing grains of melanin pigment that form characteristic black spots, 1–2 mm in diameter. The disease is not considered dangerous but badly affected fish lose much of their decorative quality. Because of the complicated life cycle of this parasite, which has to pass through the stomach of water birds and freshwater gastropod molluscs, the ailment cannot be transmitted in the aquarium from fish to fish. Infection can only come about by the introduction of untreated or inadequately quarantined molluscs into the tank.
Treatment – Eliminate possible intermediate carrier (molluscs). The black spots, in any case, will not go away.

4. Chilodonella

This is a ciliate protozoan ectoparasite (*Chilodonella cyprini* is the best known), flat–oval in shape, measuring up to 70 μm in length. It is covered with rows of cilia that slide over the epithelial cells on which it feeds, sucking out the contents. Infected fish scratch themselves and tend to keep the fins closed. Translucent whitish patches of 1–3 cm (1 in) form on the skin, the borders being very conspicuous and seen most clearly when the fish is facing the observer. If the gills are also infected, the breathing rate increases considerably and the fish spend most of their time in the upper water layers of the aquarium. In the more advanced stages, the skin may flake off. In saltwater fish the same symptoms are caused by the marine counterpart of the genus *Chilodonella*, *Brooklynella*. *Chilodonella* afflicts mainly weak or injured fish who may then transmit the disease to healthy individuals. So it tends to be most severe and widespread in crowded aquaria. Dead fish should be checked immediately for the parasites because they abandon the hosts within at most two hours after death.

The anatomy of an average aquarium fish can help in identifying the main organs and areas effected by the most common diseases. Some illnesses are caused by parasites accidentally introduced into the aquarium, while others are due to the presence of bacteria as a consequence of poor hygiene and a neglect of maintenance tasks.

Treatment – Freshwater fish, if they are not too weak, can be bathed in 3% salt solution, provided they show no signs of suffering. Alternatively, they can be bathed for 10–15 minutes in 1% salt solution. The temperature of the water can be raised to 28 – 30°C (82 – 86°F). Best results are obtained by bathing for several hours in a solution of malachite green (0.15 mg/l) or a solution of tripaflavine (acriflavine) of 1 g/100 l at a high temperature (30°C/86°F). Under these conditions *Chilodonella* dies within 10 hours. Always maintain plenty of aeration. Treatment is best carried out in quarantine tanks. In an aquarium empty of fish, at the above temperatures, *Chilodonella* disappears completely in 5–6 days for lack of hosts.

5. Cloudy eyes

This is a symptom rather than a disease proper, but it may be caused by a variety of infections. One of the most recurrent symptoms is temporary inflammation of the cornea, caused by the fish rubbing itself against some object. In other cases it may be an indication of bacterial infections or even tuberculosis. Sometimes the opacity of the eyes is caused by worm larvae (verminous cataract) or by an attack of *Oodinium* or *Cryptocaryon* (see Velvet disease and White-spot disease).
Treatment – Temporary inflammation will clear up in a few days but it may be useful to add a few drops of methylene blue to the water. In the case of bacterial infection, change the water, clean the aquarium and use general-action antibiotics because to try to identify the bacteria responsible and hence find a specific cure is both a difficult and costly exercise. One of the best remedies is oxytetracycline (250 mg/5 l for 3 days, changing the water every 24 hours to keep the antibiotic levels constant).

6. Columnaris disease

Columnaris disease, also known as mouth fungus, is caused by bacteria associated with fish mucus and known as *Flexibacter columnaris* (previously *Chondrococcus columnaris*) because the bacilli tend to appear in columns. Clinically the disease can be chronic, acute or hyperacute, its gravity depending upon the water temperature (it occurs above 18 – 20°C/64 – 68°F) and the virulence of the bacterial stock. The disease appears particularly on the head near the mouth, in the form of raised whitish, red-edged growths similar to those caused by a fungus. Lesions of the gills are necrotic and rapidly lethal. Bleeding ulcers may form on the skin. The fish draw in their fins and seem to rock in the water. The disease strikes the Poecilidae (guppies) in particular and effects apparently healthy specimens after a change of water or rise in temperature.
Columnaris is highly contagious via the water, nets and aquarium equipment, and death occurs within only 24 hours. Most vulnerable are fish with wounds or small abrasions,

a) mouth
b) eye
c) epidermis and scales
d) gills
e) liver
f) stomach
g) intestine
h) skeleton
i) swim bladder
j) fins

A

B
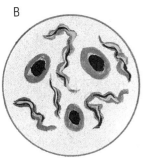    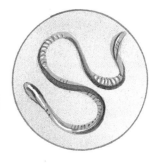

above, Section of the skin: a layer of epidermis with muciparous cells and dermis with blood vessels. below, Flagellate protozoans infect a fish's blood.

A fish's eye is easily infected by various diseases caused by parasites or following shock. The main symptom is a clouding of the eyes.

Sporozoans can multiply very rapidly by means of spores and form cysts on the gills, skin and internal organs, where they cause swelling and haemorrhaging.

Trematodes are common parasitic crustaceans that infect the gills and cells of the epidermis. The rear part is made up of a sucker with hooks.

Platyhelminths, or flatworms, distinguished by their ribbon-like body, usually infect a fish's digestive system. They can be introduced into the aquarium by means of live food.

these being the most usual lines of bacterial attack. The disease is less common in freshwater aquaria with a low pH and little organic material.
Treatment – The most effective therapy is to improve the environmental conditions as soon as possible, changing most of the water so as to reduce the amount of organic matter, increasing oxygenation and lowering the temperature to the minimum level compatible with the survival of the fish. Bathe with nifurpirinol (1 – 4 mg/l for 1 hour or 0.25 mg/l continuously), benzalkonium chloride (1 – 4 mg/l for 1 hour) or oxytetracycline (250 mg/5 l, changing the water every 24 hours for fresh with the same dosage). Alternatively, give brief baths with copper sulphate in a concentration of 1:2000 (1–2 minutes) or malachite green in a concentration of 1:15,000 (30 seconds). These same solutions can also be used to disinfect equipment after use, an essential procedure if there is more than one aquarium with no specific equipment for each one. As a preventive, you can also add a few teaspoonfuls of cooking salt to the water.

7. Copepods

Apart from copepods of the genus *Lernaea*, there are others capable of infecting aquarium fish, notably those belonging to the genera *Ergasilus* and *Caligus*. These mainly attack the gills of larger and slower-moving fish, attaching themselves by means of hook-like antennae. These parasites measure about 2 mm in length and the best known, *Ergasilus sieboldi*, is notable for its bluish coloration, very easy to identify. The most evident symptom is accelerated breathing, caused by the destruction of the branchial epithelium on which the parasites settle, resulting in hypertrophy and restriction of the blood vessels which leads to anaemia. The copepods may also be seen by raising the gill covers of diseased fish; they appear as oval spots on the gill filaments. Fungal infections may develop on the damaged gills (see *Saprolegnia*-type infections).
Treatment – Repeated baths in salt water (1% for 10 – 15 minutes), with 40% formalin (2.5 cm^3/10 l for 1 hour) or with the organophosphorus-based treatments which are available from aquarium shops.

8. Costiasis

This disease is caused by the flagellate protozoan *Costia (Ichthyobodo) necatrix*, about 10–15 µm long. In its free-swimming stage the parasite is oval in shape with two pairs of flagella, which describe rapid spiral movements in the water. In the fixation stage it is more wedge-shaped and without evident flagella. It penetrates the epidermal cells by means of a kind of hook and reproduces on the body surface of the fish, causing necroses of the cells, irritation and excessive secretion of mucus. Infected fish rub themselves against the substrate. The skin eventually develops opaque zones and then disintegrates to expose bleeding parts. The disease mainly attacks fish weakened by poor environmental conditions or excessive overcrowding.
Treatment – The best remedy is gradually to raise the temperature to 30 – 32°C (86 – 90°F), provided the fish show no signs of distress. At this temperature the parasite dies within a few days. Alternatively, bathe the fish in 3% salt water until they appear trouble-free and then place them in the quarantine tank. Protracted baths (several hours) in malachite green (0.15 mg/l) for at least a week have reliable effects. The disease can also be eradicated with tripaflavine or acriflavine in a dosage of 1 g/100 l. Do not use these when plants are present as they are harmful to them. In an aquarium without fish the parasite dies in little more than a day for lack of hosts.

9. Cryptocaryoniasis or Marine white-spot disease

This is one of the most frequently encountered diseases in tropical marine aquaria and is caused by the protozoan ciliate parasite *Cryptocaryon irritans*, which may reach 1 mm in diameter, with a horseshoe-shaped nucleus. The parasite normally infects the epidermis, fins and gills, but is sometimes found, too, on the cornea and the inside of the mouth. It looks like a white spot in the skin. When mature, it breaks the skin of the host and enters the water: it then burrows in the substrate and divides, producing up to 2,000 oval, ciliated cells which leave cysts that infect the fish. Diseased individuals are characterized by white spots over the body and fins; the skin is covered in mucus, breathing is rapid and the eyes are opaque. Initially the fish rub themselves against stones and corals. The disease is highly contagious and is spread quickly from one fish to another within only a few hours. Secondary infections of bacteria and fungi often accompany the original infection.
Treatment – The parasite is treatable only in its motile phase, before it burrows, or after the cysts burst. Raise the temperature of the aquarium to 26°C (80°F). Treat with copper sulphate (toxic to invertebrates) in the dosage of 0.15 – 0.20 mg/l in a quarantine aquarium for 3–10 days. You can also find good treatments in aquarium shops.

10. Dactylogyrosis

This disease is caused by small worms (e.g. *Dactylogyrus* spp.) which grow to 2 mm long. The parasites mainly attack the gills, destroying them as a result of continuous activity and the organs with which they adhere to the host's body. The symptoms are evident chiefly in the breathing, which is very rapid, and the gill covers, which stay raised. In the final stages, the fish remain near the surface, gasping for breath. *Dactylogyrus* attacks principally smaller fish, for adult and healthy individuals are capable of fighting it alone.
Treatment – Immersion in a 3% salt solution (for freshwater fish) gives good results, as do 30-minute baths in water and 40% formalin (using 20 – 25 cm^3/100 l). Alternatively, proprietary treatments are available from aquarium shops.

11. Dropsy

This disease may be caused by a range of factors, including bacteria and probably viruses as well. It is characterized by an accumulation of fluid in the effected organs and tissues, causing a marked swelling of the abdomen, raising of the scales and protrusion of the eyes. Small blisters may appear along the lateral line. The abdomen is filled with a reddish fluid, and the liver may be enlarged and turn green as a result of secreting too much bile. Ulcers may be evident and the diseased fish soon dies. Bad environmental conditions encourage the onset of the disease.
Treatment – Results are uncertain but it is worth trying a bath cure with nifurpirinol (1 – 4 mg/l for 1 hour or 0.25 mg/l continuously).

12. False neon fish disease

The symptoms are similar to those of neon fish disease, but the cause seems to be a bacterial infection. Here, too, the fish, mainly Characidae, grow thin and the coloration at the base of the dorsal fin becomes increasingly pale, to such an extent that the musculature is visible. Swimming is clearly difficult. The distinction between true and false neon disease is very hard to tell. Moreover, the same symptoms may be provoked by the attacks of other protozoans, such as *Chilodonella* spp.
Treatment – The principal cure is preventive, namely thorough cleaning of the aquarium. Should infection occur, treat with antibiotics such as nifurpirinol (1 – 4 mg/l for 1 hour or 0.25 mg/l continuously) and oxytetracycline (250 mg/5 l) in a quarantine tank for 3 days, changing the water every 24 hours and maintaining the same medicinal concentration to keep the antibiotic level constant.

13. Fin rot

This is a fairly common aquarium disease, due essentially to bacterial infection. The effected tissues harbour bacteria of different genera (*Aeromonas*, *Pseudomonas* and *Vibrio* in the case of marine fish), which leads to the conclusion that this ailment is caused by bad hygienic conditions of the tanks and other concomitant factors, such as a diet poor in vitamins, fins damaged by other fish or by inexperienced management of the occupants. At first, the edge of the fins is enlarged and becomes opaque; then the interradial membrane disintegrates, leaving visible the single rays, which then start to rot in their turn. As a rule, the fish dies when the fins are completely destroyed. The disease, which especially effects fish with very large, long fins, is contagious.
Treatment – Eliminate the environmental cause by thoroughly cleaning the tank and filter and changing the water. Antibacterial preparations are the best forms of medicine. Bathe the fish with oxytetracycline (250 mg/5 l for 5 days). If the infected fin area is extensive, you can try to cut out the

effected zone with a very sharp pair of scissors. Alternatively, use treatments recommended at your aquarium shop.

14. Fish pox

This disease has been known among breeding fish, such as carp, for more than 400 years, but only quite recently has it been shown to be of viral origin. It takes the form of raised warts that may appear over most of the body: initially they are grey–white but as they spread they take on a pinkish colour, due to the large network of supporting capillaries. At the start they form small whitish blisters which join together. The lesions may be long-lasting but the infection rate is low.
Treatment – There are no specific remedies. Keeping a close check on the aquarium conditions may cause the symptoms to vanish in a few months.

15. Gyrodactylosis

This disease is caused by monogenetic trematode worms belonging to genera such as *Gyrodactylus* and *Gyrodactyloides*. They measure up to 1 mm long and settle for preference on the skin, adhering to it by means of their hooked fixative organs. They feed on the epidermal cells and the mucous layer, causing the fish to exude an excessive amount of cuticular secretion, this being responsible for the opaque, whitish appearance of the entire skin surface. Other symptoms may include jagged fins, a tendency to rub against objects and the formation of ulcers.
Treatment – Immersing freshwater fish for 15 minutes (or less if it shows signs of discomfort) in a 3% salt solution is very effective. Also recommended are 30-minute baths in a well aerated solution of formalin (20 cm^3/100 l). There are specific proprietary treatments, such as Masoten; always follow instructions to the letter.

16. Ichthyosporidium disease

This is a fungus, *Ichthyosporidium hoferi*, which attacks both saltwater and freshwater fish. Internal and external organs are equally liable to infection. Externally, it appears as a sandpaper-like roughening of the skin, caused by mycotic granulomas, black in colour, about 1 mm in diameter, raised above the skin surface, particularly in the lateroventral tail region. Further development of the fungus leads to the formation of localized necrotic abscesses and ulcers. Infection of the internal organs causes the appearance of raised white nodules similar to the granulomas of tuberculosis, with which the disease may be confused. The infection is transmitted by the ingestion of infected food, and it would seem that the macroscopic lesions appear after some 30 days.
Treatment – In its advanced stage, the disease always prove lethal. In its initial phases the fish may be isolated and treated with antibiotics in a quarantine tank. If the cures have no effect, the fish should be painlessly destroyed. The aquarium and all its equipment should be disinfected with sodium hypochlorite before being reused.
Nor is there any guaranteed cure for branchiomycosis (gill fungus), although if the disease has not reached an advanced stage, it is worth trying treatment in a solution of copper sulphate in a quarantine tank.

17. Internal flagellate disease

The disease is caused by mastigophoric flagellate protozoans of the genera *Hexamita* (*H. truttae*, *H. simphysodoni*) and *Spironucleus*. They are internal parasites which attach themselves to the final tract of the digestive tube and to the gall bladder, and may also attack the circulatory system, spreading to various organs, including the liver where they cause damage. Fish effected by this disease progressively lose their appetite and colour, and swim in a clumsy manner, often settling near the floor of the tank. The faeces are white and thread-like, remaining attached to the anus. Infection of discus fish (*Symphysodon* spp.) is particularly serious, being characterized by holes in the skin. Such holes, in fact, appear on the head of these fish, and sometimes along the lateral line, discharging a whitish mass and growing increasingly large until they destroy the surrounding tissue. The wounds become sites of secondary bacterial and fungal infections.
Treatment – There is a specific medium for this disease known as Hexaex, available in some aquarist shops. Alternatively use metronidazole or dimetridazole, mixed in the food, in the proportion of 1%, every 12 hours for 3 consecutive days, or as a 1-hour bath for 3 – 5 consecutive days, in the concentration of 7 mg/l for metronidazole or 40 mg/l for dimetridazole.

18. Internal worm diseases

These infections are caused by internal parasitic worms, the most common of which belong to the groups Acanthocephala, Cestoda (tapeworms), Nematoda (threadworms) and Trematoda (flukes). The symptoms of an attack by any of these endoparasites are generally the same and they include stunted growth, a wasting body and a swollen abdomen. The most reliable diagnosis is based on examination of the faeces and the end section of the intestine, according to the following simplified procedure:
– When the fish remain motionless in the water, it is possible to see brown or red worms of 5 – 10 mm hanging from the enlarged anal opening (nematode infection of the genus *Camallanus*).
– Flat, oblong, whitish lumps, often connected to one another, present in the faeces, are visible with a microscope (cestode worm infection similar to tapeworms found in humans).
– Tapered eggs with pointed ends present in the faeces are visible with a microscope (eggs of Acanthocephala).
– Oblong eggs topped at either end by a sort of plug, similar to a bottle top, present in the faeces are visible with a microscope (infection of *Capillaria*, nematode worms).
Treatment – Curing parasites of the intestinal tract is quite difficult and is based fundamentally on administration of medicated foods. Infestation by the parasites comes from unchecked fresh or live food or as a result of introducing fish that are already infected. Scrupulous hygienic measures are generally enough to prevent the occurrence of these diseases.

19. Leeches

The most common of these parasites is *Piscicola geometra* which may appear in the aquarium along with plants or organisms gathered in the wild and not disinfected. It is recognized by two suckers, one at either end of the body, and by its jerky movements. Apart from their debilitating effect on the fish from which they suck blood, leeches are potentially dangerous as carriers of blood protozoans (see Sleeping sickness) and for secondary infections that may develop in the wounds they cause.
Treatment – Thoroughly disinfect everything that has come from natural surroundings with salt water before transferring it to the aquarium. Baths in 3% salt solution are the best means of getting rid of these parasites, and the treatment should continue until they have completely disappeared. Another possible solution is to hang a piece of raw meat, wrapped in thin netting, in the tank so as to attract the leeches and then to dispose of them by immersion in formalin, alcohol or boiling water.

20. Lymphocystis

This is a viral disease which in natural conditions has an incubation period of many weeks. It is rarely lethal. Afflicted fish initially show small pearl-shaped swellings, singly or in groups, generally on the skin, the fins or the tail, and less often on the gill plates and in the internal organs where the symptoms may be mistaken for those of white-spot disease. The cysts increase in size and become more fibrous, assuming the form of a blackberry or cauliflower, hence the vernacular name of cauliflower disease.
Treatment – There is no really effective cure. Lymphocystis is contagious so isolate the effected fish, then let the disease run its course. The cysts could be cut out, using methylene blue to clean the wound.

21. Neon fish disease

The cause of this infection is a microsporidian protozoan, *Pleistophora hyphessobryconis*, which attacks the muscles of Characidae and some Cyprinidae, forming large swollen patches and destroying them. The clearest symptoms are loss of colour at the base of the dorsal fin and loss of balance. The disease is transmitted from one fish to another when spores from a sick individual are released into the water and then ingested by another fish. The ameboid form that

appears in the digestive tract of the new host then travels through the circulatory system towards the musculature where the sporoblasts form, causing new infections (see also False neon fish disease).
Treatment – There is no really effective cure but good results may be had from baths in terramycin and aureomycin (500 mg/70 l). Very soft water with a pH between 5 and 6 prevents the onset of the disease: for this purpose peat filtration is useful. Should the aquarium become infected, it must be emptied and thoroughly disinfected.

22. Saprolegnia-type infections

The organisms responsible for these infections are aquatic fungi of the genera *Saprolegnia* and *Achyla*, distinguished from each other by the fact that the former, when out of water, look like sticky, formless masses while the latter look more compact. Clear signs of infection are the appearance of white lesions that resemble cotton wool on the skin and fins, or on the eyes, of either dead or live fish. This is often a secondary infection of already weakened individuals.
Treatment – Baths in solutions of cooking salt or of malachite green often prove useful. In some instances the effected part can be swabbed directly with tincture of iodine or 1% methylene blue; in addition to getting rid of any eggs that are affixed, the proliferation of egg fungus can be checked by using methylene blue (5 mg/l). Avoid contagion and disinfect tank with formalin.

23. Septicaemia

This form of infection is caused by bacteria belonging to the genera *Pseudomonas* and *Aeromonas*. The principal parts effected are the skin, the spleen and the kidneys. The main symptoms on the skin are large bleeding lesions and swelling of the epidermis through accumulation of fluid. This is rapidly followed by ulceration and extension of the lesions to the underlying muscles. Other symptoms may include thinning, anaemia of the gills and loss or darkening of colours.
Treatment – Infections of this type are always the result of deteriorating conditions in the aquarium environment. Healthy fish have sufficient natural defenses to fight this form of infection. In the first stages of the illness, improvement of the surroundings brings about a remission. If this is unsuccessful, try an antibiotic cure, as for example oxytetracycline (250 mg/5 l) in a quarantine tank for 3 days, changing the water every 24 hours, with another of the same medicinal concentration to maintain a constant level of the antibiotic.

24. Sleeping sickness

This disease is caused by flagellate protozoans of the genera *Cryptobia* or *Trypanosoma* which live in the blood or on the gills of effected fish, causing them to swim slowly and to lose normal flight reactions. The eyes are sunken, without reflexes. The belly is hollow and the fish tend to remain in a sideways position, head turned downward. The gills are pale. Certain diagnosis of this disease comes only from examination of the blood and sometimes the faeces with a microscope of at least 200 magnification. In the blood of fish afflicted by sleeping sickness there are flagellate organisms 15 – 25 µm in length with two flagella, one at the front and the other turned backward, provided with a waving membrane. The disease is transmitted by leeches which can be detected and eliminated.
Treatment – A cure can be attempted with parachlorophenoxethol (1 ml of solution to every 100 g of fish, added to their food), but success is uncertain.

25. Tuberculosis

This disease is caused by mycobacteria (e.g. *Mycobacterium piscium, M. marinum, M. fortuitum, M. anabanti* and *M. platypoecilus*) which attack a large number of tropical aquarium fish, especially freshwater species. The symptoms are very variable and often coincide with those of other ailments, as indicated in parentheses below. The fish, in fact, may show spinal deformations (sporozoan infections), weight loss, back in the form of a knife blade, hollow abdomen and increasingly dark coloration (intestinal parasites), protruding eyes (dropsy), open but not purulent ulcers, and slow movements. Sometimes the fish dies without any of these external symptoms being visible. The disease is transmitted by the ingestion of infected material, food containing fish viscera and, in viviparous fish, through the ovary.
Treatment – The bacteria of fish tuberculosis may be transmitted to humans, causing infections and/or hypersensitivity of the skin; so if it is suspected, it is advisable always to wear gloves and never to touch the fish with bare hands, particularly if any cuts and sores are present. There are no guaranteed cures, although a reasonable measure of success may be had by supplying fresh or live food and bathing the subjects in a solution of 2.5 g of acriflavin and 2 g of methylene blue to every 1,000 l of water for 2 weeks. Even better, however, is a bath with aureomycin (26 mg/l) for 4 days, repeating the dose on the second day of treatment.

26. Tumours

These involve an abnormal proliferation of cell tissues, manifested in the swelling and increased size of the effected organs. As occurs in humans, these may be benignant or malignant, and the cure is always uncertain, there being no proven remedies. One of the most widespread and perhaps the most easily curable form is tumour of the thyroid, which may be diagnosed by the appearance on the floor of the mouth or beneath the gill cover, which appears raised, of a reddish or pinkish swelling, a few millimeters in diameter, which tends gradually to spread.
Treatment – Administration of prolonged baths with a solution (5 mg/l) of potassium iodide dissolved in the water may cure thyroid tumours.

27. Velvet disease

The cause of this disease is the flagellate protozoan *Oodinium limneticum* or *O. pillularis*, related to the species that causes the same disease in marine fish (see Velvet disease of coral fish) and has the same life cycle. This is an external, pear-shaped parasite which attacks the gills and body and which, in the advanced stage of the disease, assumes a powdery appearance. The protozoans may pass from the gills to the digestive system where they attack and infest the intestinal mucus. When the infection is widespread, the body of the fish takes on a bright, sometimes golden, sheen. With a strong magnifying glass it is then possible to see on the epidermis or on the edges of the scales whitish or yellowish spots up to 70 µm in diameter. Initially, before these symptoms are evident, the fish show signs of restlessness and tend to rub themselves against objects. The disease sometimes appears unexpectedly when the fish, after living in equilibrium with their parasites, are suddenly weakened for one of a variety of causes, or in fish under stress after just being introduced to the aquarium. Moreover, the cysts of *Oodinium* spp. may remain quiescent in the substrate for even as much as six months before maturing and releasing the infectious flagellate spores.
Treatment – Use tripaflavine (1 g/100 l) for baths of a few hours. A useful alternative is methylene blue in doses of 6–8 drops of 1% solution to every 4 liters of water for 5 days. If this cure is repeated, leave an interval of 3 days and change the tank water completely before starting the new treatment.

28. Velvet disease of coral fish

The origin of this disease is the flagellate protozoan *Oodinium* or *Amyloodinium ocellatum*, similar to the species *O. limneticum* or *O. pillularis*, which cause freshwater velvet disease. It is an external parasite, rounded in shape, that attacks the gills and body of the fish which, in the advanced stages of the disease, looks as if it is covered with dust. On the epidermis and edges of the scales appear tiny whitish or yellowish spots up to 100 µm in diameter. At the start of the infection, the fish rub themselves against the stones and corals in the tank and breathe rapidly. The disease spreads gradually until it obstructs the gills and causes cutaneous lesions; it is then quickly transmitted from one fish to another. When the parasite is mature, it abandons the body of the fish and falls to the bottom of the tank where it forms cysts, inside which the cell divides repeatedly. Within 3 days at 25°C (77°F) a single cyst produces 259 spores, each of which is furnished with a flagellum and consequently swims off in search of a host to infect. The disease sometimes appears

unexpectedly when the fish, after living in equilibrium with their parasite, are suddenly weakened for one of a variety of causes.
Treatment – It is possible to add copper sulphate (1 – 2 mg/l for 3–10 days) directly to the aquarium, provided there are no invertebrates, for whom copper is toxic. If this is the case, it is better to treat the fish in a separate tank, also because in this way the parasites die for lack of hosts; the fish should stay on their own for some time because the flagellate spores remain active for 72 hours.

29. White-spot disease or Ichthyophthiriasis

This is the freshwater counterpart of the marine fish disease caused by *Cryptocarion irritans* (see Cryptocaryoniasis). Here the cause of the disease is the ciliate protozoan *Ichthyophthirius multifiliis*, which may measure up to 1 mm in diameter, being distinguished from *Oodinium* by its horseshoe-shaped nucleus. Normally the parasite infects the epidermis, gills and fins, but sometimes it is found too on the cornea and inside the mouth. It looks like a white spot under the skin. When mature, the protozoan ruptures the skin of the host and enters the water, causing epithelial erosion and cuticular thickening. Once free, the parasite forms cysts in the substrate and divides rapidly, producing some 2,000 oval ciliate cells which after 10–20 hours leave cysts that infect the fish. The body and fins of the effected fish are covered with tiny white spots that in the advanced stages of the disease assume the appearance of a covering of dust. There is an abundance of mucus, breathing is rapid and the eyes are opaque. Initially the fish rub themselves against the substrate. The disease is highly contagious and is quickly transmitted from one fish to another, sometimes in a few hours. Secondary infections of fungi and bacteria are often associated with this first infection.
Treatment – The parasite is sensitive to cure only in its motile phase before it forms cysts and after these burst. The remedy is to bathe the subject in malachite green (1 mg/20 l for 10 days or 0.1 g/100 l for 4–5 hours). For those Characidae (neon tetras, cardinal fish, etc.) that cannot tolerate malachite green, use baths with acriflavine (1 g/100 l). By using a quarantine tank and removing all the fish from the aquarium, you will ensure the tank is disinfected, because at temperatures above 23° – 24°C (73° – 75°F) the parasites mature within 24 hours and the new protozoans can only survive 48 hours with a lack of hosts.

PRINCIPAL CURATIVE PRODUCTS AND METHODS OF ADMINISTRATION

Acriflavine, Tripaflavine: these come in the form of orange–red tablets, due to the presence of acridine, a pigment of this colour. The product should be dissolved to make a solution of 10 mg/l at pH 7 (neutral). It is light-sensitive and must not be dissolved in salt water. It may be toxic to plants and is unsuitable for the treatment of fry and eggs.

Copper sulphate: one of the oldest remedies by reason of the disinfectant properties of copper. It is available in the form of crystals or blue powder and is mixed with water to obtain a 0.1% (1 g/1 l) stock solution. It is toxic to invertebrates and plants.

Formalin: colourless liquid with a pungent smell which is normally sold in 40% concentrated solutions. The formalin to use for therapeutic purposes should be free of methanol and kept in dark bottles or in a dark place. A whitish deposit on the bottom is indicative of an old or toxic products. Handle with care; avoid inhaling the vapour and any contact with mucous membranes.

Malachite green: available in the form of dark green or yellow–green crystals. The former are more effective but potentially toxic to fish. This product must not come into contact with zinc or products containing zinc; it deteriorates in light and is partially inactivated by organic substances. It is best to prepare each time a 0.1% (1 g/1 l) stock solution.

Methylene blue: a well-known blue colorant and disinfectant. Handle with care because it can stain the skin for lengthy periods. Buy it in a 1% solution. It is partially inactivated by organic substances and at high concentrations is toxic to plants.

Metronidazole, Dimetridazole: the preparations based on these products are used in standard medicine against flagellate parasites. For use in the aquarium, choose tablets.

Neguvon: a commercial product related to Masoten. Both are antiparasite compounds of the family of phosphoric dimethylesters available under the usual name of Trichlorfon. When using, pay special attention to the manufacturer's instructions.

Nifurpirinol: an antibacterial compound and fungicide that has the advantage of being absorbed through the gills and thus acting internally as well as externally, without having to be mixed with food.

Oxytetracycline, Terramycin, Aureomycin: antibiotics that have to be administered with extreme care. It is essential not to apply them in doses lower than prescribed or for less time than recommended, in order to avoid the development of organisms resistant to treatment, thus making it impossible to cure the fish.

Phenoxethol, Parachlorophenoxethol: oily, liquid products often sold under the commercial name Liquitox in a 1% concentration.

Potassium permanganate: available as purple crystals, best used in a 0.1% (1 g/1 l) stock solution. Becomes inactive with organic substances and when used in neutral or alkaline water (e.g. marine aquaria), tends to form a product that can damage the gills. For this reason, increase aeration.

Sodium chloride: ordinary cooking salt, freely available and thus regarded as one of the speediest and most effective preventive and remedial treatments for many freshwater fish species.

METHODS OF TREATMENT

Bathing: the fish are bathed in a tank partially containing the medicament for a duration that varies according to concentration and illness but always with abundant aeration.

Dispersal: the product is added directly to the aquarium, as, for example, a few teaspoonfuls of cooking salt or several drops of methylene blue as a mild disinfectant.

Injection: this type of treatment should be reserved for larger fish and only undertaken by experts. Injections should preferably be given in the abdomen.

Local application: this essentially comprises the application of ointment or drops to localized areas of the body. The fish must be removed from the water, placed on a moist cloth and gently held still. If the fish struggles too much, treatment is impossible and it is best to put it back in a tank on its own and not to try again until it calms down. This type of treatment has to be carried out quickly in order to avoid causing the subject too much stress.

Oral administration: the medicine is mixed with the food and given by mouth. Dry food must be moistened beforehand in the medicinal solution. Before applying this treatment, make sure that the fish are eating normally, as many diseases involve systematic refusal of food.

Prolonged immersion or long bath: the fish remain in the medicinal solution, in a quarantine tank, for a long period (one or more days).

Rapid immersion or short bath: the fish are immersed in a concentrated chemical solution for up to a few minutes.

Surgical operation: minor surgery can be carried out even by aquarists without great difficulty. The simplest procedures involve removing parts of a rotted fin and getting rid of large external parasites. Effected fish should be handled in the manner recommended above for local application of medicaments. Instruments include sharp scissors of various sizes, scalpels and tweezers, which must always be thoroughly sterilized.

Glossary

acid Water is defined as being acid if it contains an excess of hydrogen ions. Acid water is generally rich in carbon dioxide. Opposite of "alkaline."
adipose fin Fleshy fin, without rays, situated behind the dorsal fin.
adsorption Process of accumulating substances on a solid but extremely porous surface. A typical adsorbent material is activated carbon.
air lift System for circulation of water inside pipes which exploits the air and water mixture that forms as a result of blowing air into the water by means of aerators.
albino Animal without pigmentation. This characteristic may be obtained by an adaptation to life in surroundings where light is absent.
alevin Newly born fish.
alkaline Water is defined as being alkaline it if contains an excess of hydroxyl ions. Opposite to "acid."
anaesthetic Substance that decreases sensitivity and gives rise to a sleep-like state.
anal fin Fin inserted in the abdomen between the vent and the tail.
antibiotic Substance that inhibits the proliferation of, and destroys, pathogenic microorganisms. It is of great therapeutic value.
Artemia Tiny marine shrimp known as brineshrimp; both the adults and larvae (nauplii) are widely used as aquarium food.

bacteria Single-celled microorganisms which live separately or in colonies, generally reproducing by division.
barbel Fleshy appendage situated around the mouth.
basalt Dark-coloured volcanic rock.
benthos Collective description of aquatic organisms living in the depths.
bioconditioners Mixture of elements and other substances and bacteria designed to increase the speed of filter maturation.
biological equilibrium Balance obtained between organisms and the surroundings in which they live, or established within a group of organisms.
biotic Relating to living organisms.
biotope The environment in which an animal or plant population lives.
brackish Water of medium to low salinity which forms in zones where fresh and sea water mix.

carotenoids Yellow–violet pigments of animal or plant origin, very important in processes of cell metabolism.
catabolites Waste products of cell metabolism, generally made up of nitrogenous substances that have not been assimilated.
caudal fin Tail fin of fish.

caudal peduncle The part of a fish's body that precedes the caudal fin; as a rule, it is narrower than the rest of the body.
class Taxonomic category that comprises a number of associated orders.
commensalism Symbiotic relationship between two individuals of different species who use food resources in common; usually, only one of them derives the benefits, but without harming the other.
ctenoid Type of soft lamellate scale with dentate rear edge.
cycloid Type of soft, rounded lamellate scale with smooth rear edge.
Cypriniformes All fish that have some affiliation with the Cyprinidae, the family that contains carp and goldfish.
clay Compact, minute sediment that renders soil impermeable.
coral reef Typical rocky formation of tropical seas made up of the calcareous skeletons of hard corals and other colonial animals.
courtship livery Particularly bright coloration assumed by many fish during the spawning period. Sometimes it is the only method of distinguishing males and females.

density The proportion of mass to volume. Based on equal volume, sea water weighs more than fresh water.
dentate With toothed edges.
dermic Connective cutaneous tissue.
detritus Fragments of decomposing matter.
detritivore Animal that feeds on detritus.
dilution Addition of solvent to a solution to diminish its concentration.
disinfectant Substance that destroys pathogenic germs.
dorsal fins Fins situated on the back of fish.

ecosystem The unit formed by a community of animals, plants and their environment.
ectoparasite Parasite that lives outside the body of its host.
endemic Said of a species, race or organism that lives in one particular area.
endoparasite Parasite that lives inside the body of its host.
epidermis Superficial tissue covering the body of an animal.
epoxy resin Synthetic resin in the form of a sticky fluid that can be smeared, shaped and hardened with the addition of special substances. It is used by aquarists in building backgrounds and decorative fixtures.
erectile Capable of being set in raised position.
ethology Science of animal behaviour.

exophthalmus Protrusion of eyeball beyond the eye cavity.
exoskeleton Superficial skeleton composed of organic substances (such as chitin, protein), often saturated with calcareous salts.

family Taxonomic category that comprises a number of similar genera.
fertilizers Substances used for increasing soil productivity.
filiform Thin, long, thread-like.
filter Apparatus, situated inside or outside aquarium for purifying water by passing it through porous material(s).
filter base Porous plate through which the water is drawn.
filter bed The filtering medium that extracts debris from the water.
float Type of glass suitable for building aquarium tanks.
fluorescent Synonym of neon lamp, in which light is not produced by an incandescent filament but by an electric discharge created between two electrodes and which is subsequently transformed into light by the inner covering of the tube.
freeze-dried aquarium food Dried food for marine fish which reconstitutes in the water.
fry Young, newly born fish.

gastropod Mollusc with well developed feet and, as a rule, a single, conical dorsal shell.
genus Taxonomic category that comprises a number of similar species.
gills Organs for gaseous exchange, characteristic of aquatic animals.
gonopodium Copulatory organ derived from a transformation of the anal fin.

habitat Environment in which a given species lives.
herbivore A plant eater.
hydrodynamic Possessing a body shape ideal for moving in water.
hydrometer Instrument used to measure the density of a liquid.

infection Pathological condition caused by germs.
inflorescence Flowers arranged in groups.
infusorians Single-celled protozoan organisms characterized by the presence of cilia.
intertidal Coastal zone in which tidal activity occurs; it remains submerged during high tide and is exposed at low tide.
invertebrate Animal without a backbone or inner bony skeleton.
ion exchange Process of water purification that occurs thanks to particular products capable of capturing and retaining dissolved toxic substances. Very often the

substances that benefit from this principle are highly specific to a substance or group of similar substances.

larva Temporary developmental phase of an animal which has not yet assumed its definitive form nor its ability to reproduce.
lateral line Sense organ present along the sides of a fish.
limestone Rock composed of calcium carbonate.
livery Coloration of an animal.
lumen Unit of measurement of the flow of radiant energy emitted by a lamp.

madrepore Corals; marine animals with an exoskeleton which constitute a coral reef.
mangroves Aquatic plants with aerial roots, typical of tropical swamps.
mantle Cutaneous formation, typical of molluscs, which produces the shell.
metamorphosis The overall sequence of phases through which an animal evolves as it attains its definitive adult form.
methylene blue Coloured substance used for preparing microscope slides and as an antiseptic.
microorganism Organism of such a minute size that it can be observed only through a microscope.
mimicry Phenomenon whereby an organism assumes colours or forms similar to those of another species or of its environment.
monochrome Of a single colour.
morphology Science studying the form, structure and development of living organisms.
mucus Thick fluid that covers and protects the epidermis or the internal mucous membranes of an animal.

nauplius Characteristic larva of a crustacean.
nitrates Salts of nitric acid.
nitrification Chemical process whereby bacteria transform ammonia into nitrites and nitrates.
nitrites Salts of nitrous acid.

operculum Bony plate that covers the gills.
order Taxonomic category that comprises a number of similar families.
osmoregulation Control of the water content and salts concentration in the body of an animal. Each organism is suited to a determined concentration of salts and can sustain only limited variations. Fish have particularly sensitive osmoregulation systems.
oviparous Reproducing by laying eggs.
ovoviviparous Reproducing by means of eggs that develop inside the body of the female.
ozone Gas consisting of an isotope of oxygen which has oxydizing and disinfecting properties.

parasite Organism that lives at the expense of another.
pathogenic Giving rise to disease.
pathology Science engaged in the study of disease.
peat Substance that forms by plant decomposition in swampy areas.
pectoral fins Fin situated on each side of the body behind the operculum.
pelagic Said of an organism that lives in the open sea, not in the depths.
Perciformes Order including perch and fish similar in appearance.
pH Chemical symbol used to denote the alkalinity or acidity of a solution. If below 7, the solution is acid; if above 7, alkaline.
photosynthesis Process whereby plants exposed to light synthesize sugars by consuming carbon dioxide and releasing oxygen.
phylum Taxonomic category that comprises a number of similar classes.
pigment Substance giving colour to the tissues of living organisms.
plankton The living organisms and animals that exist far from the sea bed, suspended in the water. Minute in size, they provide an abundant source of food for many forms of marine life.
plate Type of crystal suitable for aquaria.
poikilotherms Organisms that assume the same temperature as their surroundings and are conditioned by it.
polychaetes Marine worms with segmented body, belonging to the Anellida.
polyp Small animal of the Coelenterata with a sac-shaped body that lives attached to the bottom in large colonies, some of which make up the calcareous structure of coral.
protein Organic substance composed mainly of molecules of oxygen, hydrogen and nitrogen; it is a fundamental constituent of living matter and plays an important part in metabolic processes.
protrusile Capable of being thrust forward.

radula Mouth organ of molluscs, composed of chitinous and renewable teeth arranged in transverse rows.
regeneration Reproduction of a part of an organism that has been detached or destroyed.
rhizome Elongated underground root containing reserve substances from which roots develop.
rock pools Pools of water that form along reefs and beaches as a result of the activity of waves and high tides and that remain isolated from the sea when the tide recedes; although they contain some water permanently, the action of tides means they are subject to variations of salinity and temperature.
root Part of a plant, generally underground, that plays a vital role in support and nutrition (absorbing substances from soil).
rotifers Tiny animals, found mainly in fresh water, characterized by mobile and retractile cilia.

salts Chemical compounds that form from the reaction between an acid and a base. They are fundamental elements for an animal and plant life and constitute a large part of rocks.
scales Small plates that cover the body of many fish.
scission Asexual reproductive process whereby an organism divides, thus generating two or more individuals similar to itself.
secretions Emission of particular substances, with various functions, from parts (especially glands) of certain organisms.
sediment Deposits of various kinds that form on seabeds.
seed Element of an individual plant or animal that enables it to reproduce.
sexual dimorphism Evident difference in appearance between male and female individuals of the same species. It may involve variation in shape, size or colour, or in a particular structure.
shoal A multitude of fish that habitually swim together.
solution Homogeneous mixture of two or more substances, of which one is usually in a liquid state.
species Taxonomic category that comprises similar individuals who may interbreed, thus generating other fertile individuals.
spermatophore Small "packet" of sperm that males of certain species transfer to the females for purposes of fertilization.
stem Part of a plant that serves to provide support and to transport nutrients; it is situated above the roots.
suboperculum Plate situated beneath the operculum.
substrate Soil in which plants and animals live; in the aquarium it is the material that makes up the tank bottom.
suspension Presence, within a liquid, of microscopic suspended particles which do not fall to the bottom.
swim bladder Bladder formed from the evagination of the intestine, which has a hydrostatic function.
swimming membrane Membrane that facilitates and permits swimming, present in many aquatic animals.
symbiosis Biological association of two or more individuals of different species, which is advantageous to all concerned.
synthetic materials Substances produced artificially by chemical processes.

taxonomy Classification system, notably of plants and animals.
territorial Typical behaviour of many animals that consists in the occupation and defense of a precisely defined area.
thermocline Zone that separates the surface waters, which are heated by the sun, from those lying deeper, which maintain a lower, constant temperature because they are not influenced by solar radiation.
trace elements Substances essential for all living forms, present in nature in very minute quantities.
tubercle Small rounded out-growth.
Tubifex Annelid worms that live in water, immersed in mud. They are used as food for aquarium species.
tunicates Marine animals that belong to the phylum Chordata and are characterized by an epidermal covering.

variety Modification of the secondary characteristics of a species.
vent Opening at the end of the digestive tract, through which waste is excreted.
ventral fin Unpaired fin situated in front of the vent, on the abdomen.
viral Caused by a virus.
virus Simple microorganism that causes numerous infections.
voltmeter Instrument for measuring electric voltage.

watt The energy given off by a lamp.
weed A spontaneous plant that invades the soil to the detriment of cultivated species.

zoophagous Feeding on animals.
zooplankton The animal organisms that make up plankton.
Zostera Marine plant that forms extensive underwater prairies.

Index

Abramites, 136
Abudefduf abdominalis, 68
Abudefduf bengalensis, 68
Abudefduf saxatilis, 68
Abudefduf troschelli, 68
Acanthophthalmus semicinctus, 38
Acanthuridae, 148
Acanthurus achilles, 68
Acanthurus leucosternon, 68
Acanthurus lineatus, 68
Acanthurus triostegus, 69
Achilles tang (*Acanthurus achilles*), 68
Acropora cervicornis, 128
Actinia equina, 128
Adionyx suborbitalis, 69
Aeoliscus strigatus, 69
Aequidens curviceps, 38
Aequidens pulcher, 38
African catfish (*Clarias batrachus*), 45
African lungfish (*Protopterus dolloi*), 57
Agassiz' dwarf cichlid (*Apistogramma agassizi*), 40
Aholehole (*Kuhlia taeniura*), 93
Alestes longipinnis, 38
Amphiprion, 131, 146
Amphiprion ephippium, 69
Amphiprion frenatus, 69
Amphiprion ocellaris, 70
Anabantidae, 141
Anabas, 141
Anabas testudineus, 38
Anablepidae, 138
Anableps anableps, 39
Anableps microlepis, 139
Anampses meleagrides, 70
Angelfish (*Pterophyllum scalare*), 57
Anisotremus taeniatus, 70
Anisotremus virginicus, 70
Anomalopidae, 143
Anomalops, 143
Anomalops katropron, 71, 143
Anoplarchus purpurescens, 114
Anoptichthys jordani, 39
Anostomidae, 39, 136
Anostomus anostomus, 39
Anostomus trimaculatus, 39
Antennariidae, 142
Antennarius hispidus, 71
Antennarius sanguineus, 71
Anthias, 113
Anthopleura elegantissima, 128
Anthopleura xanthogrammica, 128
Aphyosemion australe, 39
Aphyosemion sjoestedi, 40
Apistogramma agassizi, 40
Apodichthys flavidus, 114
Apogonidae, 144
Apogon maculatus, 71
Apogon retrosella, 72
Apolemichthys arcuatus, 72
Aponogeton crispus, 126
Arapaima gigas, 135
Arapaimidae, 135
Arbacia punctulata, 128
Archerfish (*Toxotes jaculator*), 60
Argus fish, see Scat
Argus grouper (*Cephalopholis argus*), 75
Arothron meleagris, 72
Arothron reticularis, 72

Arowana (*Osteoglossum bicirrhosum*), 54
Aspidontus taeniatus, 73, 93
Asterias forbesi, 128
Asterias rubens, 128
Astronotus ocellatus, 40
Astyanax mexicanus, 40
Atlantic porkfish (*Anisotremus taeniatus*), 70
Atlantic toadfish (*Opsanus tau*), 117
Australian lungfish (*Neoceratodus forsteri*), 53

Balanophyllia elegans, 128
Balistapus undulatus, 73
Balistes carolinensis, 114
Balistes vetula, 73
Balistidae, 149
Balistoides conspicillum, 73
Banded barb (*Barbus pentazona*), 41
Banded butterfly fish (*Chaetodon humeralis*), 78
Banded cardinalfish (*Apogon retrosella*), 72
Banded cleaner goby (*Gobiosoma digueti*), 87
Banded comber (*Serranus fasciatus*), 106
Banded pipefish (*Dunkerocampus dactyliophorus*), 85
Bandit angelfish (*Apolemichthys arcuatus*), 72
Banner blenny (*Emblemaria hypacanthus*), 85
Barbodes, 41
Barbus, 41
Barbus conchonius, 41
Barbus cumingi, 41
Barbus hexazona, 41
Barbus nigrofasciatus, 41
Barbus oligolepis, 41
Barbus pentazona, 41
Barbus schwanenfeldi, 42
Barbus tetrazona, 42
Barbus tetrazona partipentazona, 41
Barbus tetrazona tetrazona, 42
Batrachoididae, 149
Bay blenny (*Hypsoblennius gentilis*), 92
Beacon fish (*Hemigrammus ocellifer*), 49
Beaubrummel (*Stegastes flavilatus*), 106
Belontiidae, 141
Betta splendens, 42
Big-eye squirrelfish (*Myripristis murdjan*), 97
Bird wrasse (*Gomphosus varius*), 88
Bitterling (*Rhodeus sericeus*), 63
Black goby (*Gobius niger*), 116
Black molly, 56
Black platy, 61
Black ruby barb (*Barbus nigrofasciatus*), 41
Black tetra (*Gymnocorymbus ternetzi*), 48
Black triggerfish (*Melichthys niger*), 95
Black widow, see Black tetra
Black-eyed goby (*Coryphopterus nicholsi*), 115
Black-tailed humbug (*Dascyllus melanurus*), 84

Blacknosed butterfly fish (*Johnrandallia nigrirostris*), 92
Blenniidae, 148
Blennius ocellaris, 114
Blennius tentacularis, 115
Blenny (*Tripterygion nasus*), 121
Blind cavefish (*Anoptichthys jordani*), 39
Bloody frogfish (*Antennarius sanguineus*), 71
Blue angelfish (*Holocanthus bermudensis*), 90
Blue chromis (*Chromis cyanea*), 81
Blue devil (*Chrysiptera cyanea*), 82
Blue gourami, 61
Blue ribbon eel (*Rhinomuraena amboinensis*), 105
Blue-banded goby, see Catalina goby
Blue-girdled angelfish (*Euxiphipops navarchus*), 86
Blue-lined surgeon, see Clown surgeon
Blue-spotted boxfish (*Ostracion tuberculatus*), 99
Bothidae, 149
Bothus mancus, 74
Botia macracantha, 42
Brachydanio albolineatus, 43
Brachydanio rerio, 43
Brachygobius xanthozona, 43
Brachionus, 60
Braided butterfly fish (*Chaetodon paucifasciatus*), 79
Brineshrimp (*Artemia*), 158
Bronze catfish (*Corydoras aeneus*), 46
Bumblebee fish (*Brachygobius Xanthozona*), 43
Butterfly blenny (*Blennius ocellaris*), 114
Butterfly fish (*Pantodon buchholzi*), 55

Cabomba aquatica, 126
Caesio, 145
Caesionidae, 145
California scorpion fish (*Scorpaena guttata*), 119
Callichthyidae, 138
Callichthys, 138
Callichthys callichthys, 43
Callionymidae, 149
Canthigaster punctatissima, 74
Canthigaster valentini, 74
Capoeta, 41
Carassius auratus, 62
Carcharhinidae, 150
Carcharhinus leucas, 150
Cardinal tetra (*Cheirodon axelrodi*), 44
Carnegiella, 136
Carnegiella strigata, 44
Caulerpa taxifolia, 126
Centrarchidae, 140
Centriscidae, 143
Centropomidae, 139
Centropyge argi, 74
Centropyge bispinosus, 74
Centropyge flammeus, 75
Centropyge flavissimus, 75
Centropyge heraldi, 75
Centropyge kennedy, 74
Centropyge loriculus, 75
Centropyge potteri, 75

Centrostephanus, 85
Cephalopholis argus, 75
Cephalopholis miniatus, 76
Cerianthus, 128
Chaenopsidae, 148
Chaenopsis alepidota, 76
Chaetodipterus faber, 76
Chaetodipterus zonatus, 76
Chaetodon aculeatus, 76
Chaetodon argentatus, 77
Chaetodon auriga, 77
Chaetodon auriga auriga, 77
Chaetodon auriga setifer, 77
Chaetodon collare, 77
Chaetodon ephippium, 77
Chaetodon falcifer, 78, 80
Chaetodon falcula, 80
Chaetodon humeralis, 78
Chaetodon kleini, 78
Chaetodon lunula, 78
Chaetodon madagascariensis, 79
Chaetodon mertensii, 79
Chaetodon ocellatus, 79
Chaetodon ornatissimus, 79
Chaetodon paucifasciatus, 79
Chaetodon rafflesi, 79
Chaetodontidae, 146
Chaetodon tinkeri, 80
Chaetodon ulietensis, 80
Chaetodon xanthurus, 79
Chaetodontoplus mesoleucos, 80
Chalceus macrolepidotus, 44
Chanda, 139
Chanda ranga, 44
Chanda wolfii, 44
Characidae, 136
Characiformes, 136
Cheilodipterus macrodon, 80
Cheirodon axelrodi, 44, 55
Chelmon rostratus, 81
Cherubfish (*Centropyge argi*), 74
Chirolophis decoratus, 151
Chromis, 146
Chromis atrilobata, 81
Chromis caerulea, 81
Chromis cyanea, 81
Chromis limbaughi, 82
Chromis scottii, 82
Chrysiptera cyanea, 82
Cichlasoma meeki, 45
Cichlasoma octofasciatum, 45
Cichlidae, 45
Cirrhitichthys oxycephalus, 82
Cirrhitidae, 147
Clarias, 137
Clarias batrachus, 45
Clariidae, 137
Clarion angelfish (*Holacanthus clarionensis*), 91
Cleaner wrasse (*Labroides dimidiatus*), 93
Cleidopus gloriaemaris, 96
Climbing perch (*Anabas testudineus*), 38
Clinidae, 148
Clown loach (*Botia macracantha*), 42
Clown surgeon (*Acanthurus lineatus*), 68
Clown trigger (*Balistoides conspicillum*), 73
Clown wrasse (*Coris gaimardi*), 83
Cobitidae, 137
Cockscomb (*Anoplarchus purpurescens*), 114

Colossoma brachypomum, 45
Congo tetra, *see* Zaire tetra
Congridae, 142
Conus californicus, 129
Copeina, 136
Copeina arnoldi, 46
Copella, 136
Copper band butterfly fish (*Chelmon rostratus*), 81
Coral beauty (*Centropyge bispinosus*), 74
Coral hawkfish (*Cirrhitichthys oxycephalus*), 82
Coral trout (*Cephalopholis miniatus*), 76
Coris gaimardi, 83
Coris julis, 115
Cortez batfish (*Zalieutes elater*), 121
Cortez damselfish (*Stegastes rectifraenum*), 107
Cortez's angelfish (*Pomacanthus zonipectus*), 103
Cortez's rainbow wrasse (*Thalassoma lucasanum*), 108
Cortez's sand conger (*Taenioconger digueti*), 108
Corydoras, 138
Corydoras aeneus, 46
Corynactis californica, 129
Coryphopterus hyalinus, 83
Coryphopterus nicholsi, 115
Coryphopterus personatus, 83
Cottidae, 139, 150
Cottus gobio, 62
Cromileptes altivelis, 83
Crossaster papposus, 129
Ctenopoma, 141
Cucumaria miniata, 129
Cuming's barb (*Barbus cumingi*), 41
Cryptocaryon irritans, 95
Cryptocorine affinis, 126
Cryptodendrum, 146
Cypraea, 129
Cyprinidae, 136
Cypriniformes, 136
Cyprinodon macularis, 46
Cyprinodontidae, 138
Cyprinus carpio, 62

Dascyllus, 146
Dascyllus albisella, 147
Dascyllus aruanus, 83
Dascyllus melanurus, 84
Dascyllus trimaculatus, 84
Dasyatidae, 150
Dasyatis pastinaca, 115
Datnioides, 140
Datnioides microlepis, 46
Dendrochirus barberi (*D. brachypterus*), 84
Dendrochirus zebra, 84
Diadema, 85
Diadema antillarum, 129
Dibranchus erythrinus, 121
Diodon histrix, 84
Diodon holacanthus, 84
Diodontidae, 150
Dipnoi, 135
Discus (*Symphysodon discus*), 59
Dogfish, *see* Nursehound
Domino damsel (*Dascyllus trimaculatus*), 84
Dragon moray (*Muraena pardalis*), 96

Dunkerocampus dactyliophorus, 85
Dwarf rainbow cichlid (*Pelmatochromis kribensis*), 55

Earth-eater (*Geophagus jurupari*), 47
Echinodorus cordifolius, 127
Eight-banded cardinalfish (*Cheilodipterus macrodon*), 80
Electric catfish (*Malapterurus electricus*), 52
Elephant fish (*Gnathonemus petersi*), 48
Elodea canadensis, 127
Emblemaria hypacanthus, 85
Emperor angelfish (*Pomacanthus imperator*), 102
Emperor snapper (*Lutjanus sebae*), 95
Ephippididae, 145
Epinephelus dermatolepis, 85
Epinephelus guaza, 116
Equetus acuminatus, 85
Equetus lanceolatus, 85
Etroplus maculatus, 47
Eucidaris tribuloides, 129
Eupomacentros rectifraenum, 107
Euxiphipops navarchus, 86
Exallias brevis, 86
Exodon paradoxus, 47
Eyed butterfly fish (*Chaetodon ocellatus*), 79

Falcate butterfly fish (*Chaetodon falcifer*), 78
False cleaner fish (*Aspidontus taeniatus*), 73
Fine-spotted jawfish (*Opistognathus punctatus*), 98
Fingerfish (*Monodactylus argenteus*), 53
Fire clown (*Amphiprion ephippium*), 69
Firemouth cichlid (*Cichlasoma meeki*), 45
Flag cichlid (*Aequidens curviceps*), 38
Flame angelfish (*Centropyge loriculus*), 75
Flamefish (*Apogon maculatus*), 71
Flashlight fish, *see* Lanternfish
Florida seahorse (*Hippocampus hudsonius*), 116
Forcipiger flavissimus, 86
Forcipiger longirostris, 86
Four-eyed fish (*Anableps anableps*), 39
Foxface (*Lo vulpinus*), 94
Foxfish, *see* Foxface
French angelfish (*Pomacanthus paru*), 103
Frogfish (*Antennarius hispidus*), 71
Fugu, 149
Fundulus chrysotus, 47

Gambusia affinis, 62, 138
Gambusia holbrooki, 62
Garibaldi damselfish (*Hypsypops rubicundus*), 92
Gasteropelecidae, 136
Gasteropelecus, 136

Gasterosteidae, 139
Gasterosteus aculeatus, 63
Gaterinidae, 145
Geophagus jurupari, 47
Ghost fish, *see* Glass catfish
Giant Cortez jawfish (*Opistognathus rhomaleus*), 98
Giant gourami (*Osphronemus goramy*), 54
Gibbonsia elegans, 86
Glass catfish (*Kryptopterus bicirrhus*), 50
Gnathonemus petersi, 48
Goatfish (*Mulloidichthys dentatus*), 96
Gobiidae, 43, 141, 148
Gobiodon citrinus, 87
Gobiosoma digueti, 87
Gobiosoma oceanops, 87
Gobiosoma puncticulatus, 87
Gobius niger, 116
Golden ear (*Fundulus chrysotus*), 47
Golden pencil fish (*Nannostomus harrisoni*), 53
Golden platy, 61
Golden puffer (*Arothron meleagris*), 72
Golden-striped grouper (*Grammistes sexlineatus*), 88
Goldfish (*Carassius auratus*), 62
Gomphosus coeruleus, 88
Gomphosus varius, 88
Goodeidae, 139
Gramma loreto, 88
Grammidae, 145
Grammistes sexlineatus, 88
Grammistidae, 144
Graveldiver (*Scytalina cerdale*), 119
Graybar grunt (*Haemulon sexfasciatum*), 89
Green chromis (*Chromis caerulea*), 81
Green puffer fish (*Tetraodon fluviatilis*), 60
Green wrasse (*Labrus viridis*), 117
Grey angelfish (*Pomacanthus arcuatus*), 102
Grunt sculpin (*Rhamphocottus richardsoni*), 118
Guppy (*Poecilia reticulata*), 56
Gymnochanda, 139
Gymnocorymbus ternetzi, 48
Gymnomuraena zebra, 88
Gymnothorax meleagris, 89
Gymnotidae, 138
Gymnotus, 138
Gymnotus carapo, 48

Haemulidae, 145
Haemulon sexfasciatum, 89
Half-banded coolie loach (*Acanthophthalmos semicintus*), 38
Halieutaea fumosa, 121
Halieutaea retifera, 121
Hamlet (*Hypoplectrus unicolor*), 92
Haplochromis burtoni, 48
Harlequin fish (*Rasbora heteromorpha*), 57
Harlequin wrasse (*Lienardella fasciatus*), 94
Helostoma temmincki, 49
Helostomatidae, 141
Hemichromis bimaculatus, 49
Hemigrammus ocellifer, 49

Hemipteronotus pavoninus, 89
Hemitaurichthys polylepis, 89
Hemitaurichthys zoster, 89
Heniochus acuminatus, 90
Heniochus diphreutes, 90
Hermissenda crassicornis, 129
Heterostichus rostratus, 148
Hippocampus hudsonius, 116
Hippocampus kuda, 90
Histrio histrio, 90, 142
Holacanthus, 146
Holacanthus bermudensis, 90, 91
Holacanthus ciliaris, 90, 91
Holacanthus clarionensis, 91
Holacanthus isabelita, 90
Holacanthus passer, 91
Holocentridae, 143
Holocentrus rufus, 91
Hoplosternum, 138
Horned blenny (*Blennius tentacularis*), 115
Hymenocera picta, 130
Hyphessobrycon callistus, 50
Hyphessobrycon pulchripinnis, 49
Hyphessobrycon serpae, 50
Hypoplectrus cyaneus, 92
Hypoplectrus gemma, 92
Hypoplectrus gummigata, 92
Hypoplectrus guttavarius, 92
Hypoplectrus tricolor, 92
Hypoplectrus unicolor, 92
Hypostomus plectostomus, 50
Hypsoblennius gentilis, 92
Hypsoblennius gilberti, 92
Hypsoblennius jenkinsi, 92
Hypsypops rubicundus, 92

Ilyodon furcidens, 50
Indian knife fish (*Notopterus chitala*), 54
Inimicus didactylus, 107

Jack Dempsey (*Cichlasoma octofasciatum*), 45
Jackknife fish (*Equetus lanceolatus*), 85
Japanese medaka (*Oryzias latipes*), 54
Japanese pinecone fish (*Monocentris japonicus*), 96
Japanese tang (*Naso lituratus*), 97
Jewel cichlid (*Hemichromis bimaculatus*), 49
Johnrandallia nigrirostris, 92, 96
Jordania zonope, 116

King angelfish (*Holacanthus passer*), 91
Kissing gourami (*Helostoma temmincki*), 49
Klein's butterfly fish (*Chaetodon kleini*), 78
Koi carp (*Cyprinus carpio*), 62
Krib, *see* Dwarf rainbow cichlid
Kryptophanaron, 143
Kryptophanaron alfredi, 101
Kryptophanaron harveyi, 101
Kryptopterus bicirrhis, 50
Kuhlia taeniura, 93
Kuhlidae, 143

Labeo bicolor, 51
Labeo erythrurus, 51
Labridae, 147
Labrisomus xanti, 86
Labroides dimidiatus, 73, 87, 93
Labrus viridis, 117
Lactoria cornuta, 93
Lactoria fornasina, 93
Lamprologus brichardi, 51
Lanternfish (*Photoblepharon palpebratus*), 101
Lates niloticus, 139
Leaf fish (*Monocirrhus polyacanthus*), 53
Leather grouper (*Epinephelus dermatolepis*), 85
Lebiasinidae, 136
Lemon goby (*Gobiodon citrinus*), 87
Lemon tetra (*Hyphessobrycon pulchrippinis*), 49
Lemonpeel angelfish (*Centropyge flavissimus*), 75
Leopard grouper (*Cromileptes altivelis*), 83
Leopard shark (*Triakis semifasciata*), 108
Lepidosiren, 53, 135
Lepidosiren paradoxa, 51
Lepomis gibbosus, 63
Leporinus fasciatus, 51
Lienardella fasciata, 94
Limbaugh's chromis (*Chromis limbaughi*), 82
Linckia, 130
Lionfish, see Scorpion fish
Liopropoma fasciatus, 94
Liopropoma rubre, 94
Lo vulpinus, 94
Lobotes, 140
Lobotes surinamensis, 140
Lobotidae, 140
Long-horned cowfish (*Lactoria cornuta*), 93
Long-finned characin (*Alestes longipinnis*), 38
Long-nosed butterfly fish (*Forcipiger flavissimus*), 86
Long-nosed filefish, see Orange-spotted emerald filefish
Long-nosed hawkfish (*Oxycirrhites typus*), 99
Long-spined porcupine fish (*Diodon holacanthus*), 84
Longfin sculpin (*Jordania zonope*), 116
Loricaria filamentosa, 52
Loricariidae, 138
Lutjanidae, 145
Lutjanus sebae, 95
Lyretail (*Aphyosemion australe*), 39
Lysmata grahami, 130
Lysmata wurdemanni, 130
Lythrypnus dalli, 95

Macropodus opercularis, 52
Macrorhamphosidae, 143
Mailed catfish (*Callichthys callichthys*), 43
Majestic angelfish, see Blue-girdled angelfish
Malapteruridae, 137
Malapterurus electricus, 52, 137
Malapterurus microstoma, 137

Marbled hatchetfish (*Carnegiella strigata*), 44
Marbled torpedo ray (*Torpedo marmorata*), 121
Masked goby (*Coryphopterus personatus*), 83
Mediterranean grouper (*Epinephalus guaza*), 116
Meiacanthus atrodorsalis, 95
Melichthys niger, 95
Metynnis hypsauchen, 52
Micralestes interruptus, 56
Miller's thumb (*Cottus gabio*), 62
Millions fish, see Guppy
Misgurnus anguillicaudatus, 137
Mochokidae, 137
Mono, see Fingerfish
Monocentridae, 143
Monocentris japonicus, 96
Monocirrhus polyacanthus, 53
Monodactylidae, 140
Monodactylus argenteus, 53
Monodactylus sebae, 53, 140
Moon angelfish (*Pomacanthus maculosus*), 102
Moorish idol (*Zanclus canescens*), 109
Moray eel (*Muraena helena*), 117
Mormyridae, 136
Mudskipper (*Periophthalmus* sp.), 100
Mullidae, 145
Mulloidichthys dentatus, 96
Mulloidichthys martinicus, 96
Muraena pardalis, 96
Muraenidae, 142
Muraena helena, 117
Muraena lentiginosa, 96
Murex, 130
Mustelus asterias, 108
Mustelus californicus, 108
Mustelus mustelus, 108
Myriophyllum spicatum, 127
Myripristis leiognathos, 72
Myripristis murdjan, 97

Nandidae, 140
Nannostomus, 136
Nannostomus harrisoni, 53
Naso lituratus, 97
Naso unicornis, 97
Nautichthys oculofasciatus, 117
Neoceratodus, 135
Neoceratodus forsteri, 53, 135
Neoclinus blanchardi, 97
Neon goby (*Gobiosoma oceanops*), 87
Neon tetra (*Paracheirodon innesi*), 55
Nexillaris concolor, 68
Notopteridae, 135
Notopterus chitala, 54
Nursehound (*Scyliorhinus stellaris*), 119

Octopus bimaculatus, 130
Ogcocephalidae, 143
Ogcocephalus radiatus, 121
Ophioblennius steindachneri, 98
Opisthognathidae, 147
Opistognathus aurifrons 98
Opistognathus punctatus, 98
Opistognathus rhomaleus, 98
Opsanus tau, 117
Orange blenny pike (*Chaenopsis alepidota*), 76

Orange chromide (*Etroplus maculatus*), 47
Orangemice triggerfish (*Sufflamen verres*), 107
Orange-striped dottyback (*Pseudochromis flavivertex*), 103
Ornate coralfish (*Chaetodon ornatissimus*), 79
Oryzias latipes, 54
Oscar (*Astronotus ocellatus*), 40
Osphronemidae, 141
Osphronemus goramy, 54, 141
Osteoglossidae, 135
Osteoglossum bicirrhosum, 54, 135
Osteoglossum ferreirai, 135
Ostraciidae, 149
Ostracion meleagris, 99
Ostracion tuberculatus, 99
Oxycirrhites typus, 99
Oxymonacanthus longirostris, 99

Pacific brill (*Bothus mancus*), 74
Pacu (*Colossoma brachypomum*), 45
Pagurus acadianus, 130
Pakistani butterfly fish (*Chaetodon collare*), 77
Panama sergeant major (*Abudefduf troschelli*), 68
Panama-fanged blenny (*Ophioblennius steindachneri*), 98
Pantodon buchholzi, 55, 135
Pantodontidae, 135
Papiliochromis ramirezi, 55
Parablennius tentacularis, 115
Paracanthurus hepatus, 100
Paracheirodon innesi, 55
Paradise fish (*Macropodus opercularis*), 52
Paramia quinquelineata, 100
Parasicyonis, 70
Parasicyonus, 146
Pareques viola, 100
Parrotfish (*Scarus ghobban*), 105
Patira miniata, 130
Peacock wrasse (*Thalassoma pavo*), 121
Pearl danio (*Brachydanio albolineatus*), 43
Pearl gourami (*Trichogaster leeri*), 60
Pelmatochromis kribensis, 55
Pennant coralfish (*Heniochus acuminatus*), 90
Penpoint gunnel (*Apodichthys flavidus*), 114
Perca fluviatilis, 63
Perch (*Perca fluviatilis*), 63
Percidae, 139
Periophthalmus, 100
Petrolisthes, 130
Phenacogrammus interruptus, 56
Pholididae, 151
Photoblepharon, 71, 143
Photoblepharon palpebratus, 101
Physobrachia, 146
Phytichthys chirus, 151
Picasso trigger (*Rhinecanthus aculeatus*), 104
Pink-tailed characin (*Chalceus macrolepidotus*), 44
Pink-tailed triggerfish (*Xanthichthys mento*), 109
Platax orbicularis, 101
Platichthys flesus, 118

Platy (*Xiphophorus maculatus*), 61
Platypoecilus maculatus, 61
Plectorhynchidae, 145
Plectorhynchus chaetodonoides, 101
Plectropoma maculatum, 76
Pleuronectidae, 151
Pleuronichthys coenosus, 118
Plotosidae, 142
Plotosus anguillaris, 101
Poecilia latipinna, 56
Poecilia reticulata, 56
Poecilia sphenops, 56
Poecilia velifera, 56
Poeciliidae, 138
Poey's butterfly fish (*Chaetodon aculeatus*), 76
Polkadot fish, see Sweetlips
Polydactylus oligodon, 102
Polynemidae, 147
Pomacanthidae, 146
Pomacanthus, 146
Pomacanthus arcuatus, 102, 103
Pomacanthus imperator, 102
Pomacanthus maculosus, 102
Pomacanthus paru, 102, 103
Pomacanthus zonipectus, 103
Pomacentridae, 146
Pomadasyidae, 145
Popeye catalufa (*Pristigenys serrula*), 118
Powder blue surgeon (*Acanthurus leucosternon*), 68
Premnas, 146
Priacanthidae, 145
Prionurus punctatus, 103
Pristigenys serrula, 118
Pristipomatidae, 145
Prognathodes aculeatus, 76
Protopterus, 53, 135
Protopterus dolloi, 57
Pseudochromidae, 144
Pseudochromis flavivertex, 103
Pseudochromis porphyreus, 103, 104
Pseudotropheus auratus, 57
Pseudotropheus elongatus, 57
Pseudotropheus zebra, 57
Psychedelic fish (*Synchiropus picturatus*), 107
Pterois russelli (= *P. lunulata*), 84
Pterois sphex, 84
Pterois volitans, 104
Pterophyllum altum, 57
Pterophyllum dumerilii, 57
Pterophyllum scalare, 57
Pteropterus (= *Pterois*) *antennata*, 84
Purple angel, see Coral beauty
Purple chromis (*Chromis scotti*), 82
Pulple dottyback (*Pseudochromis porphyreus*), 104
Pygmy angelfish, see Cherubfish
Pygoplites diacanthus, 104
Pyramid butterfly fish (*Hemitaurichthys polylepis*), 89
Pyrrhulina, 136

Queen angelfish (*Holacanthus ciliaris*), 91
Queen triggerfish (*Balistes vetula*), 73

Racoon butterfly fish (*Chaetodon lunula*), 78

Radianthus, 70, 131, 147
Raffles butterfly fish *(Chaetodon rafflesi),* 79
Rainbow scorpion fish *(Scorpaenodes xyris),* 105
Rainbow wrasse *(Coris julis),* 115
Raja clavata, 118
Rajidae, 150
Ramirez' dwarf cichlid *(Papiliochromis ramirezi),* 55
Rasbora heteromorpha, 57
Rasbora maculata, 58
Rasbora trilineata, 58
Razorfish *(Aeoliscus strigatus),* 69
Razorfish *(Hemipteronotus pavoninus),* 89
Rectangular trigger *(Rhinecanthus rectangulus),* 105
Red crescent platy, 61
Red grouper, *see* Coral trout
Red oscar, 40
Red piranha *(Serrasalmos nattereri),* 59
Red platy, 61
Red rasbora, *see* Harlequin fish
Red saddleback clown, *see* Fire clown
Red scorpion fish *(Scorpaena scrofa),* 119
Red-headed goby *(Gobiosoma puncticulatus),* 87
Red-finned shark *(Labeo erythrurus),* 51
Red-tailed surgeon, *see* Achilles tang
Regal tang *(Paracanthurus hepatus),* 100
Reticulated puffer *(Arothron reticularis),* 72
Rhamphocottus richardsoni, 118
Rhinecanthus aculeatus, 104
Rhinecanthus rectangulus, 105
Rhinomuraena amboinensis, 105
Rhodeus sericeus, 63
Rice fish, *see* Japanese medaka
Rock umbra *(Pareques viola),* 100
Rosy barb *(Barbus conchonius),* 41
Rosy rockfish *(Sebastes rosaceus),* 120
Roundfin batfish *(Platax orbicularis),* 101
Royal gramma *(Gamma loreto),* 88
Russet angelfish *(Centropyge potteri),* 75

Saddleback butterfly fish *(Chaetodon ephippium),* 77
Saddled butterflyfish *(Chaetodon ulietensis),* 80
Sailfin molly *(Poecilia velifera),* 56
Sailfin sculpin *(Nautichthys oculofasciatus),* 117
Salmon clown *(Amphiprion perideraion),* 70
Saltwater catfish *(Plotosus anguillaris),* 101
Sandelia, 141
Sarcastic fringehead *(Neoclinus blanchardi),* 97
Sargassum fish *(Histrio histrio),* 90
Sargocentron suborbitalis, 69
Sarotherodon mariae, 58
Scaridae, 147
Scarus ghobban, 105
Scat *(Scatophagus argus),* 58
Scatophagidae, 140
Scatophagus argus, 58
Schwanenfeld's barb *(Barbus schwanenfeldi),* 42
Sciaenidae, 145
Scissors chromis *(Chromis atrilobata),* 81
Scissors-tail rasbora *(Rasbora trilineata),* 58
Scleropages, 135
Scleropages formosus, 135
Scorpaena guttata, 119
Scorpaena scrofa, 119
Scorpaenidae, 143
Scorpaenodes xyris, 105
Scorpion fish *(Pterois volitans),* 104
Scyliorhinidae, 150
Scyliorhinus stellaris, 119
Scytalina cerdale, 119
Scytalinidae, 151
Sea bass *(Serranus cabrilla),* 120
Sebastes rosaceus, 120
Sebastes rubrivinctus, 120
Sebastes serciceps, 120
Serpae tetra *(Hyphessobrycon serpae),* 50
Serranidae, 144
Serranus annularis, 106
Serranus cabrilla, 120
Serranus fasciatus, 106
Serranus tabacarius, 106
Serranus tigrinis, 106
Serrasalmus nattereri, 59
Serrasalmus niger, 59
Serrasalmus rhombus, 59
Sharpnose puffer *(Canthigaster punctatissima),* 74
Short-finned molly *(Poecilia sphenops),* 56
Shrimpfish, *see* Razorfish, 69
Siamese fighting fish *(Betta splendens),* 42
Siganidae, 149
Siganus luridus, 94
Siganus rivulatus, 94
Siluridae, 137
Siluriformes, 137
Silurus glanis, 137
Silver butterfly fish *(Chaetodon argentatus),* 77
Silver dollar *(Metynnis hypsauchen),* 52
Smallscale threadfin *(Polydactylus oligodon),* 102
South American lungfish *(Lepidosiren paradoxa),* 51
Spadefish *(Chaetodipterus zonatus),* 76
Sphaeramia nematopterus, 106
Spiny cowfish *(Lactoria cornuta),* 93
Spirobranchus giganteus, 131
Spirographis spallanzanii, 131
Splashing tetra *(Copeina arnoldi),* 46
Spotted gambusia *(Gambusia affinis),* 62
Spotted kelpfish *(Gibbonsia elegans),* 86
Spotted puffer, *see* Golden puffer
Spotted rasbora *(Rasbora maculata),* 58
Spotted rock blenny *(Exallias brevis),* 86
Squilla empusa, 131
Stegastes flavilatus, 106
Stegastes rectifraenum, 106, 107
Stenopus hispidus, 131
Stichaeidae, 151
Stoichactis, 70, 131, 147
Stonefish *(Synanceia verrucosa),* 107
Striped anostomus *(Anostomus anostomus),* 39
Striped sailfin tang *(Zebrasoma veliferum),* 109
Striped surgeon *(Acanthurus triostegus),* 69
Sufflamen verres, 107
Sweetlips *(Plectorhynchus chaetodonoides),* 101
Swissguard basslet *(Liopropoma rubre),* 94
Symphysodon discus, 59
Synanceia horrida, 107
Synanceia verrucosa, 107
Synanceiidae, 144
Synchiropus picturatus, 107
Syngnathidae, 143
Syngnathus acus, 120
Syngnathus leptorhynchus, 120
Syngnathus phlegon, 120
Syngnathus typle, 120
Synodontis, 137
Synodontis alberti, 59
Synodontis nigriventris, 59

Taenioconger digueti, 108
Tealia lofotensis, 131
Teraponidae, 144
Tetraodon fluviatilis, 60
Tetraodon palembangensis, 60
Tetraodontidae, 142, 149
Thalassoma lucasanum, 108
Thalassoma pavo, 121
Therapon jarbua, 108
Thoracocharax, 136
Thornback ray *(Raja clavata),* 118
Threadfin butterfly fish *(Chaetodon auriga),* 77
Three-spined stickleback *(Gasterosteus aculeatus),* 63
Three-spot damselfish, *see* Domino damsel
Three-spot gourami *(Trichogaster trichopterus),* 61
Three-striped humbug *(Dascyllus aruanus),* 83
Tiger barb *(Barbus tetrazona),* 42
Tigerperch *(Terapon jarbua),* 108
Tilapia *(Sarotherodon mariae),* 58
Tinker's butterfly fish *(Chaetodon tinkeri),* 80
Tobacco fish *(Serranus tabacarius),* 106
Tomato clown, *see* Fire clown
Torpedinidae, 150
Torpedo marmorata, 121
Townsend's angel, 91
Toxopneustes, 131
Toxotes, 140
Toxotes jaculator, 60
Toxotidae, 140
Triakis semifasciata, 108
Trichogaster leeri, 60
Trichogaster trichopterus, 61
Trichogaster trichopterus sumatranus, 61
Triggerfish *(Balistes carolinensis),* 114
Tripterygiidae, 150
Tripterygion nasus, 121
Tropheus moorei, 61
Turbot *(Pleuronichthys coenosus),* 118

Undulate triggerfish *(Balistapus undulatus),* 73
Unicorn surgeon *(Naso unicornis),* 97
Unio, 63
Upside-down catfish *(Synodontis nigriventris),* 59

Vallisneria spiralis, 127
Variegated angelfish *(Chaetodontoplus mesoleucos),* 80
Variola louti, 76
Velvet cichlid, *see* Oscar

Wasp goby, *see* Bumblebee fish
Whiptail catfish *(Loricaria filamentosa),* 52
White-spotted boxfish *(Ostracion meleagris),* 99
White-tip squirrelfish *(Holocentrus rufus),* 91
Wimplefish, *see* Pennant coralfish

Xanthichthys mento, 109
Xanthichthys ringens, 109
Xenomystus nigri, 54
Xiphophorus helleri, 61
Xiphophorus maculatus, 61

Yellow seahorse *(Hippocampus kuda),* 90
Yellow tang *(Zebrasoma flavescens),* 109
Yellow-headed jawfish *(Opistognathus aurifrons),* 98
Yellow-tailed surgeonfish *(Prionurus punctatus),* 103
Yellow-tailed wrasse *(Anampses meleagrides),* 70
Yelloweye rockfish *(Sebastes rubrivinctus),* 120

Zaire tetra *(Phenacogrammus interruptus),* 56
Zalieutes elater, 121
Zanclus canescens, 109
Zanclus cornutus, 148
Zebra danio *(Brachydanio rerio),* 43
Zebra Malawi cichlid *(Pseudotropheus zebra),* 57
Zebra moray *(Gymnomuraena zebra),* 88
Zebra scorpion fish *(Dendrochirus zebra),* 84
Zebrasoma flavescens, 103, 109
Zebrasoma veliferum, 109

Bibliography

L.C. Hubbs and K.F. Lagler, *Fishes of the Great Lakes Region*, Bloomfield Hills (Michigan) 1958
E.S. Herald, *Il libro dei pesci*, Milan 1962
G. Sterba, *Freshwater Fishes of the World*, New York 1963
E.M. Grant, *Guide to Fishes*, Brisbane 1965
A.J. McClane, *McClanes field guide to saltwater fishes of North America*, New York 1965
H.R. Axelrod, *Mollies in Color*, Neptune City (New Jersey) 1968
H.R. Axelrod and C.W. Emmens, *Exotic Aquarium Fishes*, Jersey City (New Jersey) 1969
W. Bridges, *The New York Aquarium Book of the Water World*, 1970
R.J. Goldstein, *Anabantoids. Gouramis and related fishes*, Neptune City (New Jersey) 1971
R.J. Goldstein, *Introduction to the cichlids*, Neptune City (New Jersey) 1971
K. Jacobs, *Livebearing aquarium fishes*, London 1971
G.R. Allen, *Anemonefishes*, Neptune City (New Jersey) 1972
F. Yasuda and Y. Hiyama, *Pacific Marine Fishes*, Neptune City (New Jersey) 1973
H.R. Axelrod, *Koi of the World*, Neptune City (New Jersey) 1973
W. Burgess and H. R. Axelrod, *Pacific Marine Fishes*, vol. 2, Neptune City (New Jersey) 1973
W. Burgess and H. R. Axelrod, *Pacific Marine Fishes*, vol. 3, Neptune City (New Jersey) 1973
R.J. Goldstein, *Cichlids of the World*, Neptune City (New Jersey) 1973
W. Wickler, *The Marine Aquarium*, Neptune City (New Jersey) 1973
W. Burgess and H.R. Axelrod, *Pacific Marine Fishes*, vol. 4, Neptune City (New Jersey) 1974
B. Walker, *Sharks and Loaches*, Neptune City (New Jersey) 1974
G.R. Allen, *Damselfishes of the South Seas*, Neptune City (New Jersey) 1975
W. Burgess and H.R. Axelrod, *Pacific Marine Fishes*, vol. 5, Neptune City (New Jersey) 1975
P. Colin, *Neon Gobies*, New Jersey 1975
F. Bianchini, S. Bruno, F. Krapp, A.C. Rossi, *Acquario*, Milan 1976
W. Burgess and H.R. Axelrod, *Pacific Marine Fishes*, vol. 6, Neptune City (New Jersey) 1976
W. Burgess and H.R. Axelrod, *Pacific Marine Fishes*, vol. 7, Neptune City (New Jersey) 1976
G. Campbell, *Saltwater Tropical Fishes in Your Home*, New York 1976
J.S. Nelson, *Fishes of the World*, New York 1976
F. De Graaf, *L'acquario marino tropicale*, Milan 1976
J. Gery, *Characoids of the World*, Neptune City (New Jersey) 1977
R.C. Steene, *Butterfly and Angelfishes of the World*, vol. 1, New York 1977
W.A. Burgess, *Butterflyfishes of the World*, Neptune City (New Jersey) 1978
W. Doak, *Fishes of the New Zealand Region*, London 1978
S.W. Tinker, *Fishes of Hawaii*, Honolulu 1978
G. R. Allen, *Butterfly and Angelfishes of the World*, vol. 2, New York 1979
R. Gannon, *Starting Right with Goldfish*, Neptune City (New Jersey) 1979
D.W. Gotshall, *Pacific Coast subtidal invertebrates*, California 1979
D.A. Thomson, L.T. Findley, A. Kerstitch, *Reef fishes of the Sea of Cortez*, New York 1979
S. Spotte, *Seawater aquariums*, New York 1979
M. Torchio, *I pesci*, Milan 1979
H.R. Axelrod, *Tropical Fish for Beginners*, Neptune City (New Jersey) 1980
S. Frank, *The illustrated encyclopedia of aquarium fish*, London 1980
R.E. Thresher, *Reef fish. Behavior and ecology of the reef and in the aquarium*, Florida 1980
D.W. Gotshall, *Pacific Coast Inshore Fishes*, California 1981
N.A. Meinkoth, *The Audubon Society Guide to North American Seashore Creatures*, New York 1981
J.E. Randall, *Underwater guide to Hawaiian reef fishes*, Kanehoe 1981
J.D. Van Ramshorst, *Aquarium of Tropical Freshwater Fish*, Lausanne 1981

A.D. Hawkins, *Aquarium systems*, London 1981
D.W. Gotshall, *Marine Animals of Baja California*, California 1982
D. Mills and G. Veevers, *The Golden Encyclopedia of Freshwater Tropical Aquarium Fishes*, New York 1982
E. Ghirardelli, *La vita nelle acque*, Turin 1982
P. Hunnam, *Acquario marino e d'acqua dolce*, Milan 1982
W. Ostermoller, *Il grande libro degli acquari*, Milan 1982
A guide to exhibit animals in the John C. Shedd Aquarium, 1983
W.N. Eschmeyer and E.S. Herald, *A field guide to Pacific Coast fishes of North America*, Boston 1983
P. Ghittino, *Tecnologia e patologia in acquacoltura*, Turin 1983
J.E. Randall, *Caribbean Reef Fishes*, Neptune City (New Jersey) 1983
V. Zupo, *Manuale pratico dell'acquario*, Florence 1984
V. Zupo, *L'acquario marino mediterraneo*, Florence 1985
H.R. Axelrod, *Atlas of freshwater aquarium fishes*, Jersey City (New Jersey) 1986
D. Mills, *Il mio acquario*, Milan 1986
W. Ladiges and D. Vogt, *Guida dei pesci d'acqua dolce d'Europa*, Padua 1988
A. Sanderse, W. Weiss, W. Neugebauer, *Il libro dell'acquario*, Padua 1988
F. de Graaf, *Enciclopedia dei pesci marini tropicali*, Milan 1989
K. Horst and H.E. Kipper, *L'acquario ottimale*, Bologna 1989
G. Brunner, *Il nuovo libro delle piante*, Milan 1989
Various authors, *Actes du IIème Congrès d'Aquariologie*, Monaco 1990
W.E. Burgess, H.R. Axelrod, R.E. Huniker, *Atlas of marine aquarium fishes*, Neptune City (New Jersey) 1990
A. Mojetta, *L'acquario*, Milan 1990
A. Mojetta, *Enciclopedia dei pesci d'acquario d'acqua dolce*, Milan 1991
R. Riedl, *Fauna e flora del Mediterraneo*, Padua 1991
G. Vevers, *I pesci d'acquario*, Milan 1991
G. Gandolfi, A. Marconato, P. Torricelli, S. Zerunian, *I pesci delle acque interne italiane*, Rome 1992
S. Spotte, *Captive seawater fishes*, New York 1992
M.R. Brittan, *A Revision of the Indo-Malayan Freshwater Fish Genus Rasbora*, Jersey City (New Jersey) n.d.
Freshwater and marine aquarium, California n.d.
E. Herald, *Fishes of North America*, New York n.d.
R.J. Kemp, *Freshwater Fishes of Texas*, n.d.

Picture sources and acknowledgements

The editor wishes to thank the following for their collaboration in providing information and illustrations:

Excalibur: 12-13, 14, 17, 18, 19a, 19l, 20, 21, 22, 23, 24, 25, 26, 27, 28, 29, 31 (W. Maggi, M. Cerri, Archivio Dennerle, Industrialfoto).
A. Ghisotti: 37, 64-65, 66b, 110-111, 112, 113, 122-123, 124b, 125.
G. Giudice: 15. A. Mancini: 124a. S. Montanari: 16, 30, 32, 33, 34-35, 132-133. Grazia Neri: 11. V. Paolillo and G. D'Amato: 36, 66a, 67, 112-113. Franca Speranza: 2, 8-9,110-111 (Lanceau-Labat/Cogis).
Dr. Flegra Bentivegna, Director of Naples Aquarium, Milan Aquarium, Acquariomania of Milan, Zooacquario of Imola, Riccardo Bedini, Vincenzo Giovanetti and Marco Orsi.